About Island Press

Island Press is the only nonprofit organization in the United States whose principal purpose is the publication of books on environmental issues and natural resource management. We provide solutions-oriented information to professionals, public officials, business and community leaders, and concerned citizens who are shaping responses to environmental problems.

In 2002, Island Press celebrates its eighteenth anniversary as the leading provider of timely and practical books that take a multidisciplinary approach to critical environmental concerns. Our growing list of titles reflects our commitment to bringing the best of an expanding body of literature to the environmental community throughout North America and the world.

Support for Island Press is provided by The Bullitt Foundation, The Mary Flagler Cary Charitable Trust, The Nathan Cummings Foundation, Geraldine R. Dodge Foundation, Doris Duke Charitable Foundation, The Charles Engelhard Foundation, The Ford Foundation, The George Gund Foundation, The Vira I. Heinz Endowment, The William and Flora Hewlett Foundation, W. Alton Jones Foundation, The Henry Luce Foundation, The John D. and Catherine T. MacArthur Foundation, The Andrew W. Mellon Foundation, The Charles Stewart Mott Foundation, The Curtis and Edith Munson Foundation, National Fish and Wildlife Foundation, The New-Land Foundation, Oak Foundation, The Overbrook Foundation, The David and Lucile Packard Foundation, The Pew Charitable Trusts, Rockefeller Brothers Fund, The Winslow Foundation, and other generous donors.

energy

energy

Science, Policy, and the Pursuit of Sustainability

Edited by
Robert Bent, Lloyd Orr, and Randall Baker

Illustrations by William Z. Shetter

Institute for Advanced Study
Indiana University

ISLAND PRESS
Washington • Covelo • London

Library of Congress Cataloging-in-Publication Data

Energy : science, policy, and the pursuit of sustainability / edited by Robert Bent, Lloyd Orr, and Randall Baker.

p. cm.

Includes bibliographical references and index.

ISBN 1-55963-910-5 (Hardcover : alk. paper) — ISBN 1-55963-911-3 (Paperback : alk. paper)

1. Power resources—Social aspects. 2. Power resources—Environmental aspects. I. Bent, Robert D. II. Orr, Lloyd. III. Baker, Randall.

TJ163.2 .E4865 2002

333.79–dc21 2002002989

British Cataloguing-in-Publication Data available.

Printed on recycled, acid-free paper

Manufactured in the United States of America

09 08 07 06 05 04 03 02 8 7 6 5 4 3 2 1

To Norman S. Care
December 20, 1937–September 4, 2001

Renowned scholar and writer,
inspiring teacher, kind and caring person.

"They are not dead who live in lives they leave behind.
In those whom they have blessed they live a life again."

—A verse sent to Eleanor Roosevelt
by a friend shortly after the death of Franklin
No Ordinary Time by Doris Kearns Goodwin

Contents

Preface

> Learning to see the world from multiple positions—if such an exercise is possible—then becomes a means to better understand how the world as a totality works.
>
> —David Harvey, *Justice, Nature and the Geography of Difference*

Energy: Science, Policy, and the Pursuit of Sustainability represents one such exercise, an attempt to look at energy from multiple disciplinary perspectives. The results of extended collaboration between colleagues at Indiana University and elsewhere, this book had perhaps humble beginnings.

Exploring the topic of global energy problems was the goal of an interdisciplinary faculty seminar established at the Institute for Advanced Study in 1996, under the direction of physics professor Robert D. Bent in his role as Newton Professor. The remit of the professorship—named for the institute's first director, Roger Newton—is to work toward the integration of knowledge, and it reflects the belief of the then-director of the institute James M. Patterson in the value of interdisciplinary discourse, the ideal of bringing together faculty in a way that transcends disciplinary biases. The seminar met regularly during a two-year period, with participants from eight different academic disciplines: physics, biology, anthropology, economics, political science, business, philosophy, and language and literature. The presentations heard and given became the nucleus for thoughts toward the integrative project, combining the natural sciences, social sciences, and humanities. This volume represents the considered results of this extended collaboration.

Appropriately, for a study that combines analysis from different knowledge bases, this volume speaks in an accessible manner, clearly laying out definitions, analysis, and the policy implications around the energy question to a broad audience. Recognizing that we cannot know for sure how long the fossil fuels we currently use as primary energy sources will last, the authors warn of resource exhaustion, of the need to plan ahead to introduce environmentally safe ways of meeting our energy needs. They invite us to become informed enough to play a significant role in formulating a sound energy and environmental policy to insure and assure the future.

The authors underline that achieving sustainability is "a complex, interdisciplinary problem that cannot be solved without a basic understanding of both the laws of nature that constrain our options and the fundamental moral, cultural, economic, and political principles that determine how humans behave." This is, indeed, a timely message that needs to be shared with a larger audience. We at the Institute for Advanced Study are pleased to have been able to support the interdisciplinary work underlying this project and to see the results now moving beyond the university.

MARY ELLEN BROWN
Director, Indiana University Institute for Advanced Study
November 6, 2001

Foreword

LEE H. HAMILTON

In a 1977 speech on energy policy, I observed that, "Although many Americans refuse to believe it, there is a serious and continuing energy problem in this country." I also noted that "a frustrating struggle with energy questions was a distinguishing feature of the 94th Congress." Sadly, I could make the same statements about the current energy situation in America and about nearly every other Congress of recent decades. Despite the oil shocks of the 1970s, the Persian Gulf war, and numerous other energy-related problems that have faced the country, America has failed to develop a comprehensive, effective, and sustainable energy policy. Establishing such a policy is as important today as ever.

Energy is the lifeblood of our society and economy. We need it to cook, to heat and cool our homes, to travel, and to work. Moreover, we are accustomed to buying energy cheaply and using it lavishly. While Americans make up less than 5 percent of the world's total population, we consume 25 percent of the world's energy.

The insatiable American appetite for fossil fuels creates numerous problems: it increases pollution, contributes to global warming, subjects us to major price fluctuations, and makes us dependent on imports from other countries. Furthermore, it is not sustainable. Although many predictions of imminent limits to oil production have been proven wrong, the supply of fossil fuels that are accessible at a reasonable cost will run out in the not too distant future, and we will be forced to turn to other energy sources.

Our dependence on imported oil complicates U.S. foreign policy. For instance, it has a major influence on American policy in the Middle East. Since we do not have nearly enough oil within our borders to meet our demand, we rely heavily on imported oil from the Middle East, which is home to roughly two-thirds of the world's proven oil reserves. This is why every U.S. president refers to the American interest in access to Middle Eastern oil as vital. That vital interest forces us to look at issues that are difficult enough on their own — such as the Israeli–Palestinian conflict, our policy toward Iraq, and our complex relationship with Saudi Arabia — through the lens of our oil needs. Were it not for our dependence on Middle East oil, we might not have gone to war with Iraq in 1991, and we would not have to spend billions of dollars each year maintaining a major military presence in the Persian Gulf.

What should be the goal of our national energy policy? In my view, it should be to provide an adequate, secure, and sustainable supply of energy at reasonable prices. This goal, of course, is much easier to state than to accomplish because some aspects of it often conflict with others. For instance, our desire to have an adequate energy supply leads us to import huge quantities of oil, but the more oil we import, the less secure our energy supply becomes. Similarly, our wish to purchase energy at reasonable prices leads us to use the most inexpensive energy sources — fossil fuels — but their use at today's levels is not sustainable for very long.

Strong political leadership is needed to reconcile these conflicting goals. Several presidents, from Jimmy Carter to George W. Bush, have made energy policy a high priority, and so have many members of Congress. But our political leaders have failed to forge a national consensus around a comprehensive approach to our energy problems. Some, such as President Carter, have emphasized conservation and investment in alternative fuels. Others, such as President Bush, have emphasized increased domestic production. None of them have provided the leadership necessary to achieve all of our energy goals.

Our failure to establish a comprehensive and effective energy policy is not a failure of our imagination or capacity, but of our political system. Many politicians are reluctant to exert the kind of leadership we need to set our energy policy on a sustainable course. A sustainable energy policy requires us to make some sacrifices, but politicians do not like to ask their constituents to make sacrifices, as Randall Baker points out in his chapter on the politics of energy. Politicians prefer to offer their constituents rewards. So when it comes time to vote on energy taxes, the short-term appeal of lower taxes outweighs the long-term benefit of conservation in the minds of many politicians, and they vote for lower tax rates. The same cost-benefit analysis applies to congressional votes on global warming, which will affect future generations far more than ours.

If our national leaders explained the importance of developing a sustainable energy policy, and pushed hard for one, I believe the American people would follow their leadership. Most Americans are keenly aware of our energy challenges and are prepared to make sacrifices to ensure that we have an adequate, secure, sustainable, and affordable energy supply.

The outline for a sound and sustainable energy policy is clear. As the editors of this book argue, a comprehensive plan must include the diversification of sources of energy, more efficient energy use, greater energy conservation, and population stabilization overseas. Increased production in the U.S. should be part of the equation, but new investment and production should be concentrated in nuclear power and renewable energy sources, such as solar and wind, rather than fossil fuels. We should also invest heavily in research and new technology that will boost the energy efficiency of automobiles and appliances, and we should find ways we can reduce our energy use. Greater investment in mass transit, for instance, could go a long way toward slashing our consumption of oil. Additionally, we should support reproductive health, education, and economic development in poor countries, which will help reduce worldwide population growth rates, and thereby limit the growth in global energy demand.

For thirty years, our nation has struggled to develop a sustainable energy policy. Until we do so, our tremendous dependence on imported energy will continue to cause us numerous problems. This unique book, which looks at America's energy situation from a wide range of perspectives, should renew our commitment to develop the comprehensive energy policy we need to maintain our way of life, preserve the environment, and provide security for future generations.

Acknowledgments

Though the roots of this book extend in many directions, the original seed was an undergraduate interdisciplinary course on *Global Energy Problems: Science and Policy.* It was developed for the first time at Indiana University-Bloomington in the spring of 1994 and taught jointly by professors Robert Bent of the Department of Physics and Laura Strohm of the School of Public and Environmental Affairs (SPEA). When Dr. Strohm left Indiana University the following year for the Monterey Institute of International Studies, a similar course entitled *Energy: Science and Policy* was taught for another two years by Dr. Bent and Dr. Randall Baker of SPEA. Following Dr. Bent's retirement in the summer of 1995, the project continued in a different guise as an interdisciplinary faculty seminar on global energy problems sponsored by the Indiana University Institute for Advanced Study. The seminar participants included, in addition to the contributors to this volume, Ben Brabson (physics), Lynton Keith Caldwell (SPEA), Dennis Conway (geography), Bruce Douglas (geology), Marcia Donnerstein (psychology), Nives Dolsak (SPEA), George Ewing (chemistry), Michael Kaganovich (economics), Rebecca Barthelmie (geography), Michele Moody-Adams (philosophy), James Patterson (business and director of the Institute for Advanced Study), Jean Patterson (SPEA), Anthony San Pietro (biology), Carol Polsgrove (journalism), Sara Pryor (geography), J.C. Randolph (SPEA), James Riley (history), Hans Peter Schmid (geography), Albert Ruesink (biology), and York Willbern (political science). We are indebted to all these people for their interest in the seminar, for their help in getting it started and sustaining it for two years, and for the breadth they added to the seminar discussions.

Besides creating the illustrations, William Z. Shetter was a valuable consultant throughout the writing stages of this book, in particular with regard to making it more accessible to lay readers. Ivona Hedin (assistant director of the Institute for Advanced Study) provided efficient and thoughtful administrative help throughout the course of the seminars. We thank Elyse Abraham Bobbitt for her improvements in writing style before the manuscripts were sent to Island Press, and Mary Ellen Brown (current director of the Institute for Advanced Study) for supporting this editorial work and for her general encouragement. We extend a special thanks to Todd Baldwin of Island Press for his many critical and constructive editorial comments and suggestions throughout the planning and writing stages of this book. His contributions have resulted in a much improved final work.

energy

Introduction

The Energy-Environment Problem

Everything that happens in both the living and the nonliving world is due to the flow and transformation of energy. Energy drives the economy, and all living creatures require it. Indeed, from a thermodynamic perspective, humans are just very complex organisms for processing energy. There can be no more fundamental question than fueling our existence.

What we often think of as development has been a process of moving out of the struggle to secure food (basic energy) for our subsistence to creating an energy surplus by harnessing the power of animals, wind, water, and other local resources. With the Industrial Revolution, the increase in productivity derived from mechanical energy is now heavily dependent on nonrenewable energy resources. Currently, most energy use in the developed world is for purposes other than basic subsistence.

There is growing concern in all nations about the long-term sustainability of the energy-intensive lifestyle that the industrialized world has developed and, moreover, whether the earth can ever support this level of development for the majority of the world's people. These concerns stem from the pressures of continuing growth in population and in energy use per capita on a planet that has finite resources and a finite capacity to assimilate wastes.

Experts do not agree on how many people the earth can support for an indefinite period, but a commonly quoted range is somewhere between 4 and 16 billion. The true number will depend on the quality of life that future generations are willing to accept and on unforeseeable technological change. How to stabilize human numbers and natural resource consumption at levels that are compatible with the earth's long-term carrying capacity—and human aspirations—is one of the most challenging problems facing humankind. Rich countries have stable or declining populations as a consequence—directly or indirectly—of greater prosperity and security. The same path to population stabilization may not be feasible or desirable in poor countries because of limited energy resources and environmental constraints. Approaches that combine prosperity with reductions in fertility rates by other means and over a shorter time span are more likely to be successful for developing nations.

World population is now at 6 billion and growing at the rate of about 1.5 percent a year, adding around 80 million people (roughly one-third of the population of the United States) to the earth annually. If this growth rate were to continue for another fifty years, world population would reach 13 billion in the year 2050. United Nations projections, under reasonable assumptions of reduced fertility, are that world population could reach between 8 billion and 12 billion in the year 2050. While some industrialized countries are currently experiencing low population growth rates—and a few, actual population decline—this has little impact on global trends because these countries hold less than 15 percent of world population. Over 95 percent of population growth is occurring in the world's least developed countries.

To see the absurdity of unlimited population growth, consider that if world population were to continue to grow indefinitely at the current rate of about 1.5 percent a year (corresponding to a doubling time of forty-seven years), the average population density on all continents of the earth, including Antarctica, would reach one person per square meter in 676 years![1] Clearly, life-support systems would collapse long before this.

World economies have grown even faster than population, averaging 3.7 percent growth per year from 1950 to 1997. The World Energy Council projections for economic growth rates in the next few decades are 2.4 percent a year for developed countries and 4.6 percent a year for developing countries, with a total average of 3.3 percent a year. The global demand for energy, the essential engine of economic development, is expected to grow at similar rates. Physical growth rates such as these, corresponding to doubling times of fifteen to twenty-nine years and an average of twenty-one years, cannot continue indefinitely on a planet with finite space and resources.

Growth in world energy demand of 3.3 percent per year could not be met for long. A steady growth rate of 3.5 percent per year, corresponding to a doubling time of twenty years, would result in five doublings in one hundred years and ten doublings in two hundred years, corresponding to increases in energy demand by factors of 32 and 1,024. Meeting such large increases in energy demand in environmentally acceptable ways would not be feasible with existing or foreseeable technologies.

The question is not whether there *will* be limits on specific physical growth rates of this sort. It is, rather, what will be the *nature* of these limits and their consequences for ecological systems and human well-being?

The world is already beginning to feel the effects of the finiteness of oil and

natural gas resources. It is projected by the American Association of Petroleum Geologists that global production of oil—currently the world's largest energy source—will peak early in the twenty-first century and decline permanently thereafter and that oil reserves recoverable economically by existing and foreseeable technologies will be close to exhaustion by 2100. Natural gas will not last much longer, because the amounts of energy stored in the earth as oil and natural gas before extraction began in the middle of the nineteenth century were about equal. To date, we have used less natural gas than oil, but the rate of natural gas use is increasing as oil reserves decline. Coal reserves are large, but there may be severe environmental constraints on the rate of coal use.

Many earlier predictions of oil and other resource production peaks and exhaustion times have been proven wrong because of unforeseen discoveries and technological developments, which during the last two centuries have always been more rapid than resource depletion. But this pattern cannot be expected to continue indefinitely—there are limits to how long nonrenewable resources can last. In the case of oil, the limit will be reached when the energy required to recover a gallon of oil is greater than the energy content of the oil. It may become economical to switch to substitutes long before that point is reached. Renewable substitutes, such as solar and wind energy, have effectively unlimited lifetimes but are nevertheless limited sources of energy because they are so widely distributed and low in concentration compared to fossil and nuclear energy. As a result, they require very large collection areas (the equivalent of one-third of the state of Wyoming to supply the current U.S. total energy needs).

Energy and Sustainability

Sustainability is necessarily a vague term, and the definition is not always clear. The most widely accepted general definition of sustainable development is that given by the United Nations' World Commission on Environmental Development:

> Development that meets the needs of the present without compromising the ability of future generations to meet their own needs.[2]

The physicist Murray Gell-Mann, Nobel laureate and author of *The Quark and the Jaguar,* offers the following definition of sustainability:

> The achievement of quality of human life and of the state of the biosphere that is not purchased mainly at the expense of the future. It encompasses survival of a measure of human cultural diversity and also of many of the organisms with which we share the planet, as well as the ecological communities that they form.[3]

An economist's definition of sustainability is given in Chapter 6:

> The preservation for future generations of a set of economic and social opportunities that are at least as rich and diverse as our own. It is not a specific *goal* so much as it is a *process* of continuous change and adaptation.

Many other similar definitions of sustainability exist that vary somewhat depending on what it is one wants to sustain. The focus of this book is *energy sustainability,* but since everything that happens in the world—all life and physical processes—involves the flow and transformation of energy, energy sustainability clearly is at the core of all sustainability issues; it is a necessary (though not sufficient) condition for sustainability in its broadest sense.

Overview

Given the critical, broad, and challenging issues discussed above, we set out in this book to contribute to the definition, analysis, and policy implications of the sustainable energy challenge in a manner comprehensible to those who are not specialists in the relevant disciplines.

As we might expect with such fundamental issues, they do not reside easily within any one of the disciplinary boxes into which the sum of human knowledge has been divided. Although the individual chapters have been written by specialists and practitioners in a variety of academic fields—some of which are not always immediately associated in everyone's mind with the "energy problem"—this work has developed from its inception within a strongly interdisciplinary framework, thus—we hope—avoiding some of the worst aspects of edited studies in terms of style, homogeneity of purpose, and integration.

Resisting the normal temptation to think of the natural sciences as the "hard" sciences and the social sciences as the "soft" sciences, we examine the uncertainties of forecasting anticipated life spans of fossil fuels. We then ask why the warnings of science are often ignored and why people behave as though some miracle cure will always spring from the unlimited ingenuity of humankind. Why—given the history of building modern society on depletable assets and a historically rapid expansion of consumption—do we not have adequate policies that recognize this condition and steer us toward an alternative resource base and consumption pattern?

Humankind's unwillingness to respond to the warnings of resource exhaustion is partly due to a history of predictions about the life spans of certain strategic minerals and other resources that proved to be wrong. For example, copper, instead of disappearing in a welter of stratospheric prices as predicted several decades ago by Meadows and Meadows in "Limits to Growth," [4] is now at historically low prices because technology (plastics in plumbing, fiber optics instead of electrical cables, and satellites) has rendered it obsolete for some of its historically most important purposes. Technology was the joker in the pack, and so there is a reasonably well founded tendency to sit back and have faith in the technology gods.

Meeting world energy needs in the twenty-first century is only *half* of the "energy problem." The other half is finding ways to do this in environmentally acceptable ways. As world oil and natural gas resources become exhausted (or too scarce to be economical), we will be forced to turn to alternatives. Coal and nuclear energy are two short-term possibilities, though both have serious environmental impacts. A more environmentally friendly long-term solution to the energy problem will require greatly expanded development of *renewable* energy sources—primarily solar and wind—coupled with worldwide improvements in energy efficiency and reduction of energy waste.

We attempt throughout this book to use the terms *efficiency* and *waste* consistently. In the thermodynamic sense, *efficiency* is defined as the fraction of energy input that is converted into useful work. *Waste* refers to energy lost due to extravagant lifestyle choices—for example, driving a 14 mpg vehicle to work instead of using a 50 mpg vehicle or taking public transportation, walking, or riding a bicycle when the latter are feasible and preferable from an energy and environmental standpoint.

This book's most important mission is to provide the reader with the range of material needed for an informed policy perspective. Of course, most people will immediately respond by saying that they do not make policy or that they are just some "small cog in a gigantic machine." However, the message of the past fifty years has been that policy comes from a groundswell of popular concern and anxiety. Rachel Carson did not create the U.S. Environmental Protection Agency, but she did express the real concerns of many people. Unfortunately, it all too often takes, in addition, some form of "crisis" to make change happen, and this usually leads to less than optimal, often ill-conceived short-run policy actions.

The chapters concern themselves with the two dimensions of the energy sustainability problem: the physical dimension and the human dimension. Chapters 1 to 3 deal with the *physical dimension* of the energy sustainability problem: the physical laws of nature that humans must live by and the daunting challenges of finding ways to provide the world with the energy it needs to sustain and advance human well-being worldwide while simultaneously dealing with the environmental consequences that threaten human health and ecosystems. Chapters 4 to 7 deal with the *human dimension:* the psychological and cultural factors that determine how we use energy, the political and economic factors that determine its governance in a democratic society, the limits of markets in responding to environmental and long-term energy problems, and the ethical problem of motivating people to protect future generations. Although there are no absolute laws such as those in the physical sciences that govern human and social behavior, there are effective laws of human nature that limit what people are willing and able to do in specific personal and cultural situations. These limits can be just as constraining as the laws of science.

Although all life processes and every action, large or small, involves energy, most nonscientists have only a vague idea of what energy is in the scientific sense and little understanding of how energy flow controls and constrains everything that happens. In Chapter 1, Robert Bent, Andrew Bacher, and Ian Thomas examine what energy is and the basic principles of energy transformations—the first and second laws of thermodynamics—which are the "rules of the game" humans must play by. Chapter 1 also includes a discussion of the insidious nature of exponential physical growth, which is unsustainable in the long term for any given material resource.

Growth of both population and energy use per capita is the basis for increases in nonrenewable resource depletion and environmental degradation.

In Chapter 2, John Sheffield examines world population trends and forecasts, projected future world energy demand, and the "realistic" options the world has for meeting this demand. How many people the earth can sustain depends in complex ways both on the type and level of lifestyle supported and on future technological developments—both of which are largely unpredictable. Population and resource projections are highly uncertain because no one knows what future generations will want, what choices they will make or, technologically speaking, what they will be able to do that might seem miraculous to us now. It is clear, however, even with these uncertainties, that meeting future world energy needs will require population stabilization, new energy resources and technologies, substantial improvements in energy efficiency, and energy conservation through changes in lifestyles.

Paradoxically, energy use contributes simultaneously to human well-being and to environmental problems that threaten the quality of human life. In Chapter 3, Russell Lee addresses the second half of this dilemma—the impacts of energy use on human life-support systems, both physical and ecological. All sources of energy have environmental impacts. Meeting present and future world energy demand without doing intolerable and irreparable damage to the environment is one of the most difficult economic, technological, political, and social problems facing humankind today. In countries at the top end of the income scale, the combination of affluence, cultural factors, and cheap energy leads to high per capita energy use. At the bottom end, poverty blocks alternatives to antiquated technologies that are wasteful of energy and harmful to the environment. An additional difficulty is the great uncertainty about how nature will react to the environmental pressures being put upon her by humankind.

It's not just population that matters; it's the number of people times the energy and resource consumption per person that is a measure of humans' impact on the environment. Affluence reduces the number of people who can be supported by a given resource and technology base. Understanding human behavior is crucial to developing realistic policies aimed at conserving energy. The social and cultural factors that shape consumption and resource use patterns in both developed and developing countries are examined by Richard Wilk in Chapter 4.

In a democratic society, policy reflects the prevailing values, desired futures, and desired quality of life of the voting public. In Chapter 5, Randall Baker focuses on the difficulties of formulating energy policies for the long run in a democratic society, in particular the need for widespread public understanding of the issues, the role of crises in mobilizing public concern and political action, and the problem of the public perception of possible crises and the links between energy and the environment when the signals and evidence are confusing.

Given resource limitations (scarcity), how those resources can be effectively and equitably allocated within the current generation and between current and future generations is the central question underlying sustainable energy. To formulate an effective sustainability policy, a basic understanding of how modern economic systems work and how they may fail to meet the needs of the present and future generations must be developed. To deal effectively with the failure of markets to fully reflect the social value of environmental resources and the needs of future generations and to formulate effective energy and environmental policies, a thorough understanding of the interactions of policies with the interconnected system of markets is essential for long-term success. For the economics of the environment and sustainable energy, the economy must be seen within the context of ecological systems, the fundamental laws of energy transformations, and the potential for continuous adaptation and technological development. Lloyd Orr addresses these fundamental economic questions in Chapter 6.

The concept of sustainability involves looking ahead to the future and therefore necessarily includes how we view our responsibilities to future generations. Whether we owe future people anything is a fundamental moral question. We presume in this book that we owe them *something*—for example, a healthy state of the biosphere and economic and social opportunities at least as rich and diverse as our own. We assume that morality requires this. There is widespread agreement on this but also disagreement about the nature and seriousness of the problem and the best way to approach it. Social inaction is more a matter of not knowing than of not caring.

Exactly what morality requires of us regarding the rights of future generations is a philosophical problem in its own right but one that is not taken up in this book. Instead, in Chapter 7, Norman Care turns to the perhaps more controversial question of what moves people to do what morality requires. While there may be fairly universal agreement that morality requires something of us regarding future people, there is no tight connection between what morality requires and actually being moved to *do* what morality requires. Care considers possible motivators for current citizens of a free society to support moral policies that honor our legacy to future generations.

We conclude with some general thoughts on the goals of this book. Energy sustainability is fundamentally an interdisciplinary problem. Our purpose is to add breadth and perspective to our understanding of this problem by approaching it from multiple vantage points and showing how the different disciplinary approaches are interrelated. We focus on basic facts and fundamental principles in the belief that a broader understanding of these principles by policy makers and the general public is needed in a democratic society to move us toward a sustainable future.

Notes

1. Albert A. Bartlett, "Forgotten Fundamentals of the Energy Crisis," *American Journal of Physics,* 46 (1978): 876–88.
2. Brundtland Report, United Nations' World Commission on Environment and Development (WCED). 1987. *Our Common Future.* Oxford: Oxford University Press, p.43.
3. Murray Gell-Mann, *The Quark and the Jaguar: Adventures in the Simple and the Complex.* New York, NY: W. H. Freeman and Co., 1994.
4. Donella H. Meadows et al., *The Limits to Growth.* New York: Universe Books, 1972.

Chapter 1

Rules of the Game

ROBERT BENT, ANDREW BACHER,
AND IAN THOMAS

We begin with the "rules of the game"—the fundamental laws of nature that govern all energy transformations. Understanding these laws is crucial to achieving sustainability--nature cannot be fooled! The two most important natural laws governing energy transformations are known in science as the first and second laws of thermodynamics, and they have profound implications regarding the sustainable use of finite energy resources. We also consider in Chapter 1 the implications of steady growth in world energy use, in particular so-called exponential growth (constant percentage growth per year) in a solar system of finite energy resources. This is not sustainable. These basic principles underlie everything else in the book.

—Editors' note

Scientific theories are often expressed in terms of "laws of nature" that describe how the world works. Although new research continues to provide scientists with a deeper understanding of the universe and occasionally forces them to modify their theories, the laws of nature are absolute laws—they cannot be changed or circumvented by human ingenuity or by technological advances. In this sense, they are the rules of the game—the "game" of human survival and well-being. We have no choice but to obey these rules, so we must strive to understand them and to find ways to live in harmony with them.

Scientists have elaborated the laws of nature as they pertain to energy. The first of these is known as the first law of thermodynamics. It asserts that although energy can be converted from one form to another, the total amount of energy in all forms stays the same (is conserved) in all physical processes—energy is never created or destroyed. This is also known as the principle of conservation of energy. A naive interpretation of the first law might lead one to conclude that, since energy is never destroyed, there is no energy crisis and we have nothing to worry about. However, as we explain later, the situation is not so simple as that. Although energy is never destroyed, some energy is dissipated—degraded into a less useful, lower-grade form—during every energy conversion. Because energy is dissipated in all interactions, and because it is essentially impossible to convert low-grade energy back into a useful, high-grade form, the total amount of useful energy in the universe is continually declining. This universal trend is equivalent to a transition from a state of order to a state of disorder—also known as the law of entropy—and is mandated by the laws of probability, specifically by the second law of thermodynamics. Put another way, a "disordered" universe is far more probable than an "ordered" one, so there is an irreversible tendency toward a disordered universe full of useless, low-grade energy.

According to our scientific theories, then, all useful energy will eventually be used up. However, the time scale for the degradation of all of the energy in the universe is immense and, in the meantime, the earth receives an abundance of useful, high-grade energy from the sun. Humans could, theoretically speaking, survive on this planet for billions of years to come by learning how to rely on the sun's energy as other earthly life forms do. However, we are running out of time to make the transition from the depletable stocks of energy sequestered over geologic eras within the earth to a reliance on the flow of energy from the sun—the ultimate source of so-called renewable energy. If we do not act reasonably soon, this inevitable transition is likely to cause immense human suffering.

In the final section of this chapter, we discuss the nature and implications of exponential (constant percentage) growth, the conflict between this kind of physical growth and sustainability, and the need to understand the arithmetic of exponential growth in order to anticipate and prepare for limits to steady growth of this kind in world population and resource consumption.

The possibility of new energy sources on earth that we are not using today is discussed in Appendix 1. Energy units, factors for converting one unit to another, and a graphical comparison of unit sizes are given in Appendix 2.

What Is Energy?

Everything that happens in the universe involves a flow and transformation of energy. Whenever a living thing or an inanimate object experiences any kind of change, energy moves from one place to another and changes form. But what is energy?

A popular (and quite accurate) conception of energy is that it is a resource that makes life easier for us—a resource that takes us from one place to another, provides heat and light, powers our entertainment devices and labor-saving appliances, and improves our quality of life. The development of energy technologies began in the middle of the eighteenth century with the Industrial Revolution, which resulted in a steadily increasing usage of energy—mainly from fossil fuels—in industry, commerce, agriculture, transportation, and the home. This in turn resulted in the steadily increasing productivity that is such a desired feature of modern economies. Two and a half centuries later, human beings are now prodigious users of energy compared to other species, using ten to a hundred times as much as is needed for biological survival.

Even though people know roughly what they mean when they talk about energy, there has always been something a little mysterious about it. It is an abstract quantity, an attribute or property of matter that cannot be seen or touched like material objects. Energy comes in many different forms, and there is a mathematical formula for computing each one.[1] It is not surprising, then, that it took scientists most of the nineteenth century to develop an understanding of energy and to discover its important operating principle: the conservation of energy.

To understand just what energy is, it is useful to look first at the many forms in which it comes. Primary energy resources on the earth include fossil fuels, natural nuclear sources, and renewable forms of energy, such as solar, wind, hydropower, geothermal, and biomass. In principle, fossil fuels such as oil, gas, and coal are renewable but only on a geological time scale—hundreds of millions of years. The nuclear fuels, deuterium and uranium, were made during the creation of the universe (the Big Bang) and in the interior of stars on a cosmological time scale—billions of years. Therefore, fossil and nuclear energy sources are fundamentally limited and depletable (see Appendix 1).[2] As these sources become depleted, humans will be forced to learn how to live on renewable energy—primarily on energy from the sun—as other species do.

Renewable energies include hydroelectric power generation, solar thermal energy, the direct conversion of solar energy to electrical energy (photovoltaic energy), wind energy, the capturing of the sun's energy in biomass, ocean thermal energy conversion, wave energy, geothermal energy, and tidal energy. Of these, only geothermal energy and tidal energy are of nonsolar origin; the others are indirect ways of harnessing the sun's radiation. Because solar radiation is a product of nuclear reactions in the core of the sun and geothermal energy is produced by the decay of radioactive nuclei beneath the earth's surface, only tidal and wave energy are of nonnuclear origin. Biomass is our source of food energy and, in fossil form, our main source of nonrenewable energy. Chapter 2 discusses renewable energies in more detail.

Clearly, all of these physical resources contain energy, and an interesting question to ask is, "How much energy do they contain?" The easiest way to think about measuring energy is in terms of what it can do. For example, a certain quantity of energy is required to boil a kettle of water that is initially at room temperature. The exact amount may vary depending on the efficiency of the kettle, but in a carefully controlled experiment, the energy required to boil a certain quantity of water can be measured very accurately. This energy could come directly from wood or natural gas burning in a stove, or it could come from electricity that was generated by wind turbines, nuclear reactions, or coal fires, for example. Whatever primary source was used, the quantity of energy needed to boil the water must have been extracted from that source and transformed—eventually into heat.

The idea that energy can be quantified and that measurable quantities of energy can be transformed from one form to another is central to an understanding of energy. A lump of coal or uranium contains a certain amount of useful energy. Once that energy has been taken—transformed into electricity or heat—the coal (or the uranium) does not have it anymore. Instead, the energy is stored in an electrical circuit or in the boiling water. In the same way, an hour of sunshine contains a certain amount of energy, which can be extracted with a solar panel or simply allowed to be absorbed by the earth.

The process of converting energy from one form to another is sometimes described as "work." *Work* and *energy* are examples of words that have precise, scientific meanings in addition to their more general meanings in ordinary language. If someone is described as having a lot of energy, this means the person is capable of doing a lot of work. In science, work and energy are related in a similar way: the amount of work that can be done by a person or a machine is a quantitative measure of how much energy that person or machine possesses.

Work is what you do when you move things around against resisting forces, such as friction or gravity—for example, when you push a lawn mower or carry a piano up a flight of stairs. The amount of work done depends on how hard you have to push or pull the object to make it move and on how far the object moves as it is being pushed or pulled. The first figure in Box 1.1 illustrates this idea with the example of a fairground attendant cranking a cable that pulls a car full of children to the top of a track. The attendant must do a certain amount of work, using stored biochemical energy in the body's muscles (similar to the stored energy in coal or wood), to pull the car up the slope against the force of gravity.

In a scientific sense, work requires motion, and no work is done if there is no motion. This statement implies that weightlifters must do work (and expend energy from their muscles) to raise a barbell, but they do not need to do any work (or expend any energy) to hold the barbell stationary over their head, even

Box 1.1. Different Forms of Energy

Work. In a certain fairground attraction, a car full of children is hoisted to the top of one arm of a U-shaped track and is then released to oscillate back and forth on the track. To start with, a fairground attendant (or an electric motor) must do work to hoist the car to the top of the track. The work done depends on the force required to pull the car full of children up the track and the distance it is pulled.

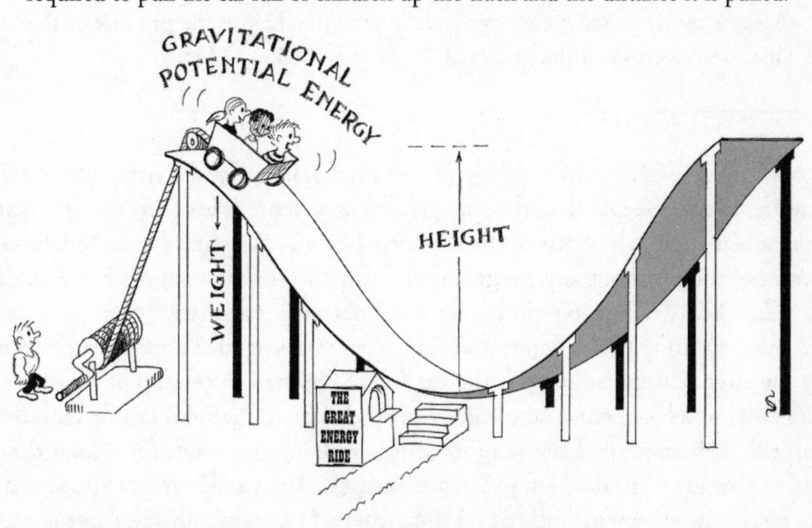

Potential Energy. Energy is transferred whenever work is done. In this case, work done by the attendant is transferred to the car full of children in the form of gravitational potential energy—energy it possesses by virtue of its position high above the ground. The gravitational potential energy (PE) of any object is defined as the product of the object's weight (its mass, m, multiplied by a gravity constant, g) and its height (h): PE = mg x h.

(*continues*)

Box 1.1. Continued

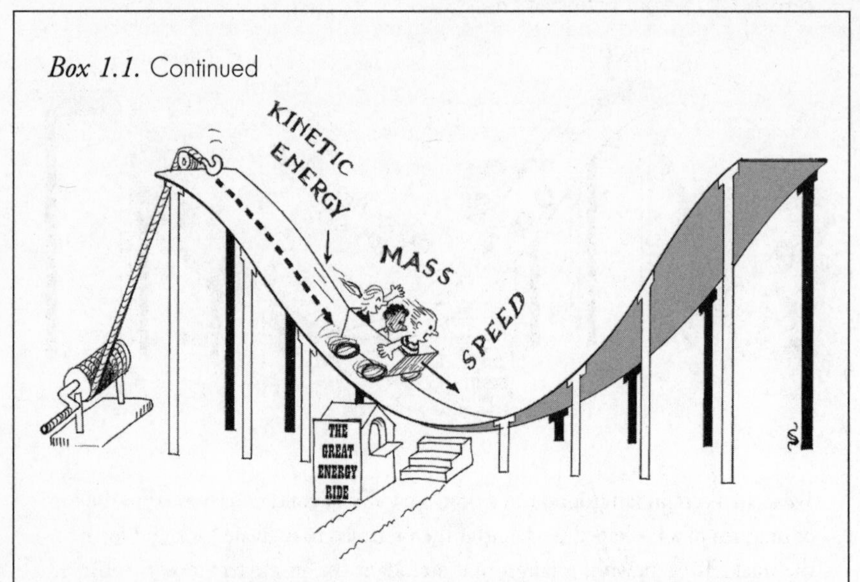

Kinetic Energy. When the brake is released, the car starts to roll down the track, gathering kinetic energy, a form of energy that depends on how fast an object is moving. Kinetic energy is defined as one-half of the product of the object's mass (m) and the square of its speed (v): $KE = 1/2\ mv^2$.

though this "effort" may make them feel tired. The scientific truth that work requires motion seems to contradict general experience. However, the fact that the barbell could, alternatively, be supported by a metal stand for an indefinite period of time without any energy expenditure shows that energy is not needed to hold a heavy object high in the air if the object is stationary.

But, returning to the fairground ride, what happens to the energy expended by the attendant when he pulls the car full of children to the top of the slope? Similarly, what happens to the energy expended by the weightlifter to raise the barbell? If these individuals were to continue doing this work for a long time without eating a meal (taking in more energy), they would simply run out of energy, so the energy must be going somewhere. In fact, as illustrated in the second example in Box 1.1, the attendant is transferring this energy to the car full of children in the form of *potential energy* (gravitational potential energy, to be exact). Similarly, the barbell acquires its potential energy from the weightlifter.

The internal structure of the car and children at the top of the track is the same as it was when they were on the ground; the only thing that has changed is their position—they have moved up in the world. *Gravitational potential energy* is defined as the energy possessed by an object due to its position in a gravitational field. The more work the attendant does, the higher the car rises and the more gravitational potential energy it acquires. Water backed up behind a dam is another example of stored gravitational potential energy. These energies of position have the potential to be transformed into useful work or other forms of energy.

Energy stored in fossil fuels and nuclear fuels is a form of chemical and nuclear potential energy, respectively, which depends on the relative positions of atoms in molecules (chemical energy) and protons and neutrons in atomic nuclei (nuclear energy), bound together by either electrical force fields or nuclear force fields.

Potential energy can be transformed into energy of motion, known as *kinetic energy.* The kinetic energy of a moving object depends on how much matter it contains and how fast it is traveling, so a high-speed tractor-trailer has a lot more kinetic energy than a slow-moving bicycle. A stationary object has no kinetic energy. In the third example in Box 1.1, the car full of children, initially stationary at the top of the track with no kinetic energy, when released starts to roll down the slope, picking up speed and, therefore, kinetic energy. By the time the car reaches the bottom of the slope, it is moving quite rapidly and has acquired a lot of kinetic energy.

The concept of kinetic energy is central to understanding why energy is so important to modern humans: we need energy to move ourselves and other things around. Most of the icons of the twentieth century—cars, trucks, and planes; agricultural equipment; and industrial machines—consume energy (much of it originating in fossil fuels) and convert it into kinetic energy to make something move.

So far, we have discussed physical stores and sources of energy, such as fossil fuels, nuclear sources, the wind, and the sun. We have also shown how the energy in these physical sources can be converted into more conceptual forms, such as potential energy (energy of position) and kinetic energy (energy of motion). The final form of energy we will consider is heat.

Historically, heat and motion were viewed as separate areas of physics. However, in the 1840s, the English physicist James Prescott Joule demonstrated the connection between heat and mechanical (kinetic) energy by measuring the rise in temperature of a liquid when it was stirred by a paddle wheel. Because of Joule's discovery, heat is now recognized as another form of energy that can be generated *from* other forms of energy (including chemical, potential, and kinetic) and converted *into* other forms. An example of the partial conversion of heat energy into mechanical (kinetic) energy is the steam engine or the internal combustion engine. The efficiency of a heat engine is defined as the fraction of the input heat energy that is converted into mechanical energy. Since the heat is usually produced by burning coal, gasoline, or some other kind of fuel that must be paid for, heat engines are designed to have the greatest possible efficiency. Although their efficiency has greatly increased since the early steam engine, it is impossible (according to the second law of thermodynamics) to make a perfect heat engine—that is, an engine with 100 percent efficiency. Some of the fuel's energy must be discharged as waste heat. The efficiencies of modern steam engines are typically 30 to 40 percent. The efficiency of an ordinary automobile engine is 20 to 30 percent, and that of a large diesel oil engine is about 40 percent.[3]

Interestingly, when heat energy is examined at the molecular or atomic scale, it turns out to be a form of kinetic energy—energy of random motion of molecules and atoms. The scientific study of heat and the transformation of mechanical energy into the random motion of molecules and atoms is known as *thermodynamics,* and the random movements and vibrations are often referred to as *internal energy* or *thermal energy.* Since thermal energy is nothing more than the kinetic energy of atoms and molecules, thermodynamics connects the macroscopic (human-scale) and microscopic (atomic-scale) domains of nature. Like heat, sound energy also involves molecular vibrations, which spread out and die away rapidly as the vibrational energy is distributed and diluted among increasing numbers of molecules.

A crucially important characteristic of heat (which we will consider in more detail in a later section) is its tendency to flow spontaneously from a hot region to a cooler region.

Joule's work on the equivalence of heat and mechanical energy (for which his name was given to the scientific unit of energy) laid the foundation for understanding an immensely powerful law of nature: the conservation of energy, known to scientists as the first law of thermodynamics.

The Conservation of Energy

The principle of the conservation of energy states that, in any physical process, energy may be transformed from one form to another but cannot be created or destroyed. Since energy can never be created or destroyed, the sum total of all forms of energy in an isolated system (one in which there is no energy transfer into or out of the system) must stay the same throughout any process. Energy is *conserved.*

The importance of the principle of conservation of energy cannot be overstated. Much of nineteenth-century science was devoted to establishing this fundamental principle. It is a generalization of experience that has not been contradicted by any observation of nature, and it is enormously useful in analyzing energy transformations.

The fairground attraction described earlier provides a simple illustration of the conservation of energy. In Box 1.2, the car full of children starts at rest at the highest point of the track, where it has gravitational potential energy but no kinetic energy. When it is released, it rolls down the track, picking up speed (and

Box 1.2. The Conservation of Energy

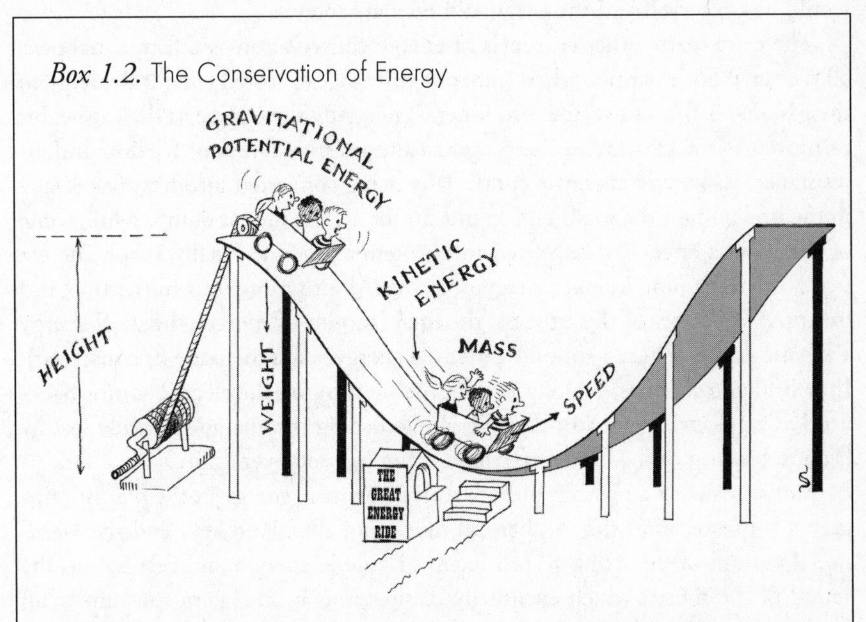

When energy is transferred from one form to another, the total amount of energy is conserved throughout the process. As the car rolls down the track, it loses gravitational potential energy (PE) but gains speed and, thus, kinetic energy (KE). The car reaches its maximum speed and kinetic energy as it passes through the lowest point on the track. At this moment, the car's gravitational potential energy is zero because its height above the lowest point is zero. The principle of conservation of energy states that the total amount of energy remains the same throughout the ride. This means the car's total energy at the highest point (purely PE) is equal to its total energy at the lowest point (purely KE). Similarly, when the car reaches the highest point on the other side and stops momentarily, all of the KE it possessed at the lowest point has been transferred back to PE. In the absence of friction, the car would continue to oscillate indefinitely from side to side, transferring energy back and forth between PE and KE with the total amount remaining constant: PE + KE = constant.

kinetic energy) but losing potential energy. At the lowest point, all of the potential energy has been converted to kinetic energy (ignoring friction) and the car reaches its maximum speed. As the car climbs the other side of the track, all of the kinetic energy is converted back to potential energy. Even though the individual values of potential and kinetic energy vary, the sum total of the two forms of energy is constant throughout this process (see Box 1.2). In the absence of friction and other resistive forces, the total energy (potential and kinetic energy) would be conserved and the car would oscillate forever.

There are many other examples of energy conservation—in fact, it happens all the time! For example, when someone drives a car, stored chemical energy in the gasoline is first converted into kinetic energy by the engine as the car begins to move. As the car travels along against the resistive forces of friction and air resistance, its kinetic energy is continually being converted into heat (especially in the tires and on the road) and sound (in the air). If the car climbs a hill, some of the kinetic energy is converted into potential energy. Finally, when the car stops, the remaining kinetic energy is converted into more heat in the tires and brakes. At all stages of this process, the total amount of energy (chemical energy + kinetic energy + heat + sound + potential energy) stays the same (is conserved). In hybrid automobiles, braking is achieved by running the electric motor backward as an electric generator and thereby returning the energy normally lost by friction braking to the batteries, where it can be used over again.

Similarly, when a piano is played, stored chemical energy in the pianist's fingers is converted into the mechanical motion of the piano keys and the vibrational motion of the strings (both forms of kinetic energy) and finally into the sound of the music, which eventually dissipates as heat. Again, the sum of all forms of energy remains constant.

So, how does the principle of conservation of energy relate to our long-term energy needs? At first glance, it might appear that there is nothing to worry about. Since energy can never be destroyed, we cannot possibly use up all energy resources, even if we wanted to. Unfortunately, the laws of nature are not as simple as that. Although energy is never destroyed, it is continually degraded.

The Irreversible Degradation of Energy

An interesting feature of the two real-world examples discussed in the previous section (driving a car and playing a piano) is that all of the energy ends up as heat or sound (which eventually ends up also as heat). In fact, all real-world examples of energy transformations (meaning that real resistive forces, such as friction and air resistance, are taken into account) share this feature. On the contrary, in the idealized example illustrated in Box 1.2, resistive forces are intentionally neglected and the car oscillates forever—its energy is continually transformed between kinetic and potential forms but never dissipated as heat. The fact that real systems always dissipate some energy is crucial for a complete understanding of the issue of energy sustainability and the fundamental role of energy in all questions of sustainability.

Heat is remarkably different from other forms of energy. Whereas other forms (such as chemical, kinetic, and potential energy) can be converted from one to another completely and each of these other forms can also be converted into heat completely, heat energy can be converted back into the other, more useful forms of energy only partially.[4] In fact, any process that converts heat back into the other forms can be achieved only by degrading a larger quantity of useful energy into heat at the same time. So, every energy transformation in which some energy is converted into heat results in a permanent loss of useful energy.

Returning to the earlier examples, there is no way to recover the heat generated during a car trip and put it back in the gas tank, and there is no way to recover the sound energy from the piano music (which is ultimately converted to heat) and return it to the pianist. Heat is referred to as *low-grade* energy, while chemical, kinetic, and potential forms of energy are known as *high-grade* energy. In Box 1.3, we introduce reality into the example of the fairground ride. Friction and air resistance cause some of the car's high-grade energy to be transformed into low-grade heat. With each swing, the maximum height reached by the car decreases.

The distinction between high-grade and low-grade energy can be illustrated with an analogy from the world of finance. One million dollars in a single lump sum is "high-grade money" because useful things can be done with it. However, $1 million distributed equally among the 6 billion people of the world would amount to less than two hundredths of a cent per person—a useless amount. Even though there would still be $1 million altogether, the process of distribution would have converted it to "low-grade money" and it would be essentially impossible ever to regain its usefulness.

The $1 million analogy provides a big clue to why heat energy is so different from high-grade forms of energy. Just as the $1 million is money shared among 6 billion people, heat energy is energy shared among billions of trillions (10^{24}) of molecules. The effect of this sharing is to dilute the heat energy's usefulness and convert it to a low-grade form. Whereas chemical, nuclear, kinetic, and potential energy are naturally contained within an object, in which they remain in a concentrated form, the nature of heat (and sound) is to disperse itself—for example, the heat from a fire naturally radiates throughout the room, the sound of thunder can be heard for miles around, and the vibrations from an

Box 1.3. The Degradation of Energy

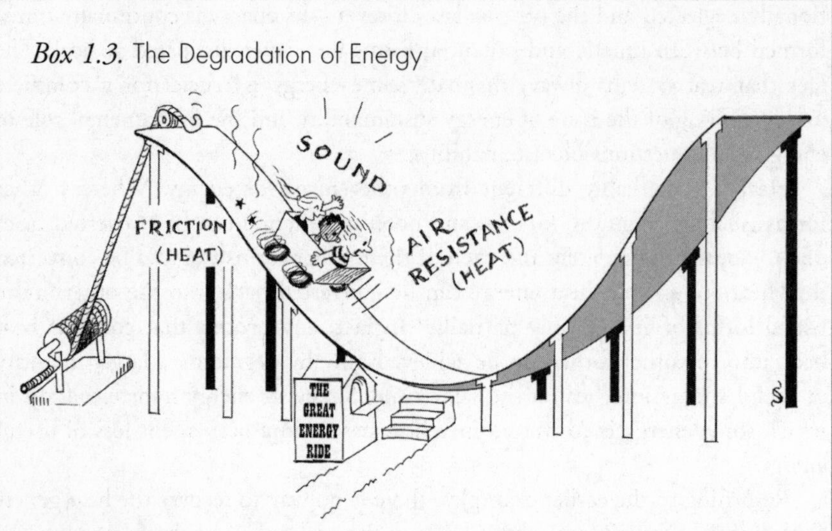

Although energy is conserved throughout any process of energy transfers, some of the energy involved in a real process is degraded into a useless form and is effectively lost. After several swings of the children's fairground ride, the car fails to rise all the way to the top of the track and hence has less potential energy than it had at the start. Since the kinetic energy is always zero at the end of a complete swing because the car's speed is zero, the car must have less total mechanical energy than it had at the start. As the car swings back and forth, mechanical energy is converted to a low-grade form and is dissipated into the earth and atmosphere. Low-grade forms of energy include heat and sound, and the processes that degrade energy into these forms include friction on the track and air resistance.

earthquake (another form of low-grade energy) can sometimes be felt thousands of miles from the source. Although it is possible to contain heat (with a thermos flask, for example), once it has been dispersed into the atmosphere or the ground, it can never be recovered.

A familiar example of irreversible energy degradation is that a hot substance is cooled down when a cold substance is mixed with it (for example, when cold cream is poured into a cup of hot coffee). The temperature of a substance is a measure of the average kinetic energy of its molecules, and in this example, the fast-moving, high–kinetic energy molecules of the hot substance transfer some of their kinetic energy to the slower-moving, low–kinetic energy molecules of the cold substance as a result of collisions. This phenomenon is similar to the way a fast-moving billiard ball loses some of its kinetic energy when it collides with a slower (or stationary) ball. When hot and cold substances are mixed, the molecules of the cold substance gain in kinetic energy (start moving faster) while those of the hot substance lose kinetic energy (move more slowly) until the temperatures of the two substances are the same. In practice, once the "hot" molecules have lost their high-grade kinetic energy, the only way to recover and concentrate the energy again is to employ a process that works by degrading a larger quantity of high-grade energy elsewhere. Alternatively, some new high-grade energy could be supplied.

In Box 1.4, when all of the car's energy has been dissipated into low-grade thermal energy and it comes to rest at the lowest point on the track, the only way to give the children another ride is for the attendant to convert more high-grade stored energy from his muscles into potential energy by hoisting the car up again.

Box 1.4. The Irreversibility of Energy Degradation

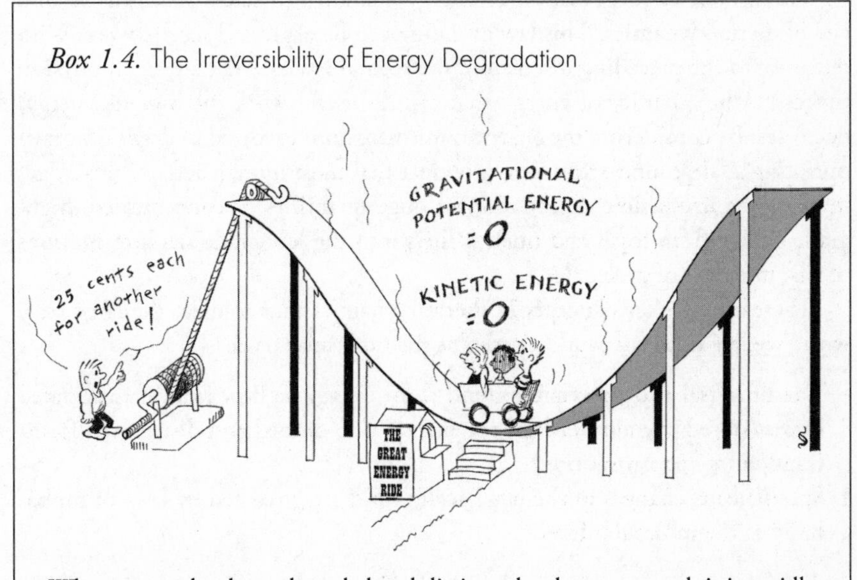

When energy has been degraded and dissipated as heat or sound, it is rapidly distributed into the earth and atmosphere, and there is no way for the car to recover it. This figure shows the children's fairground ride after a large number of swings, when the car has finally come to rest at the lowest point on the track. At this time, all of the energy originally transferred to the car by the attendant has been dissipated. It is impossible to recover the dissipated energy, so the only way to set the car in motion again is for the attendant to crank it back up to the top of the track. This means the attendant must do work to transfer more high-grade stored energy from his muscles to the car.

This cycle cannot be repeated indefinitely unless the attendant replenishes his stored energy by eating a meal. In any isolated system, there is a limited amount of high-grade energy, and when it has all been degraded, there is no more.

Of course, real fairground attractions of this kind are powered by electric motors that consume large quantities of high-grade electrical energy (much of it originating as fossil fuels) to hoist the car up. However, the same principle applies—if there is a power outage, effectively making the fairground attraction an isolated system, it will be forced to shut down.

We can extend this analogy to the entire universe, with dramatic implications. Since the universe is, by definition, "everything," it must be an isolated system—there is nothing else outside of it! Since it is an isolated system, we have to accept that it contains a limited amount of high-grade energy and will eventually run down.

The irreversible degradation of energy is known to scientists as the second law of thermodynamics. This law of nature can be expressed several ways. With reference to the preceding discussion, the second law asserts that, in all physical processes, the quantity of energy that can do useful work (high-grade energy) decreases. By considering the energy transformations involved in driving a car or operating a fairground attraction, it is clear that most human activity these days involves the irreversible degradation of huge quantities of concentrated, high-grade energy from fossil and nuclear fuels into the low-grade random motions of vast numbers of molecules.

There are two key concepts in thermodynamics that connect the large-scale world we live in to the world as seen at the molecular level:

- The universal and irreversible trend is for energy to flow from concentrated forms (stored chemical energy, for example) to diluted and distributed forms (random molecular motion).
- Spontaneous changes in the large-scale world are governed by laws of probability at the molecular level.

This section has been concerned with the first of these concepts—irreversible degradation of energy. In the next section, we will examine the concept of probability and use it to explain why the universe behaves this way.

Probability Theory: Why Energy Degradation Is Irreversible

In a certain sense, we are all familiar with the second law of thermodynamics because we encounter it all the time. It defines the direction in which changes take place and distinguishes the "possible" from the "impossible." We would be very surprised to see a stirred cup of coffee with cream spontaneously "unmix" itself so that the cold cream separated itself from the hot coffee. We would suspect trickery if a stationary fairground car extracted energy from its surroundings and started to oscillate back and forth on its U-shaped track. We expect a dropped egg to break when it hits the floor, but we do not expect the smashed shell and contents to reassemble themselves into an intact egg.

Action movies played backward are always good for laughs—especially with children—because we know how things normally behave and we recognize at once the impossibility of these reverse processes. The reverse processes are impossible because they all violate the second law of thermodynamics. Why does the second law forbid these reverse processes and impose a unique direction of change? What prevents a smashed egg from reassembling itself or a stationary fairground car from extracting energy from the air and starting to oscillate? Before answering these questions, it is important to note that these reverse energy conversions would not violate the principle of the conservation of energy.

The answers to these questions are to be found in the laws of probability. Probability and uncertainty in the microscopic world of atoms and molecules are at the heart of the second law of thermodynamics. Because of our incomplete knowledge (our uncertainty) about the exact behavior of individual molecules, the best we can do is describe molecular motion in terms of probabilities—for example, "There is a 50 percent probability that, at a particular moment, a certain number of molecules are moving in a particular direction with a particular speed." Based on these probabilities in the microscopic (molecular) world, the second law makes definite predictions about the way things behave in the macroscopic (human-sized) world—for example, the fact that low-grade heat energy cannot be converted completely back into useful, high-grade forms of energy. Making the connection between the microscopic and macroscopic worlds was one of the great triumphs of late-nineteenth-century science. The connection was made by applying the laws of mechanics statistically using formal and abstract mathematical techniques. This approach, developed by J. Willard Gibbs and Ludwig Boltzmann, among others, is called *statistical mechanics,* a term that includes the kinetic theory of gases developed earlier by Robert Boyle, Daniel Bernoulli, James Prescott Joule, A Kronig, Rudolph Clausius, and James Clerk Maxwell, among others, as a subbranch.[5]

The approach is quite subtle. We start by considering a system of ten coins on a tray, which at first sight appears to have little to do with molecules, heat, or energy conversions.[6] If the tray is shaken vigorously so that all ten coins jump in the air and spin around an unknown number of times, since there are two different ways each coin can land there are 2^{10} (or 1,024) different ways all ten coins can land back on the tray. Only one of these ways corresponds to all heads or all tails, whereas there are 252 different ways of getting five heads and five tails and 210 ways of getting six heads and four tails or four heads and six tails.

Therefore, it is much more probable that the coins will land with roughly equal numbers of heads and tails than with a very unequal distribution. For example, one-fourth of the shakes (252 out of 1,024) will yield five heads and five tails, whereas all heads or all tails will come up, on the average, only once in 1,024 shakes. Approximately equal distributions are more likely than very unequal ones because there are more ways to achieve an approximately equal distribution. The greater the number of coins, the less likely it is that the result of a single shake will be very different from an equal number of heads and tails.

Box 1.5. The "Certainty" of Equal Distributions

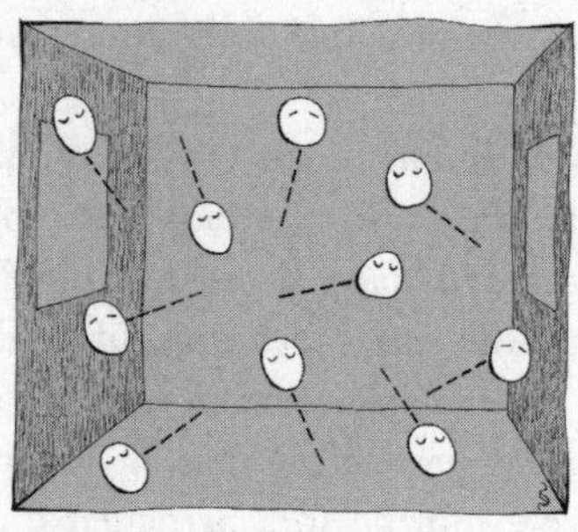

A room containing just ten gas molecules is divided into two equal parts by an imaginary wall. If the molecules are moving around randomly, each molecule has an equal chance to be on either side of the room and the mathematics for determining the distribution of the molecules in the room is the same as for the ten coins on a tray. Using the numbers calculated for the coins on a tray, there will be five molecules on each side of the room, or six molecules on one side and four on the other, about three-quarters of the time. Even so, all ten molecules will be on the same side of the room occasionally—the probability for this arrangement is 1/1024 for each side of the room, so for about four seconds every hour (on average), all ten molecules will be on the left side and for another four seconds they will all be on the right side.

In reality, there are about 10^{24} gas molecules in a typical room. With so many molecules, the probability distribution is much more strongly biased toward equal numbers of molecules on each side. In fact, the probability of the distribution deviating from equality by even one part in a trillion is infinitesimal. The probability of all 10^{24} molecules being on one side is so phenomenally small its occurrence is essentially impossible.

The role of probability in the microscopic world is illustrated in Box 1.5. Each molecule has an equal chance of being in either half of the room. With just ten randomly moving molecules in the room, the probability of all ten molecules being on one side is the same as the probability of ten coins all landing heads up—one chance in 1,024. There are actually about a million million million million (1,000,000,000,000,000,000,000,000 = 10^{24}) molecules of air in a typical room. With this enormous number of molecules, it becomes quite meaningless to talk about the probability of all of them being on one side of the room simultaneously—an event so improbable as to be essentially impossible.

The conclusion from these two examples is that when large numbers of microscopic elements (coins, for example) behave randomly, the macroscopic process (the overall distribution of heads and tails) is biased toward "average behavior" rather than "eccentric behavior." Because the numbers of microscopic elements (atoms and molecules) typically involved in macroscopic physical processes is so enormous, the bias toward average behavior, such as approximately equal numbers of molecules on both sides of a room, is quite overwhelming. Eccentric behavior, such as all the molecules in a room being on the same side, is essentially impossible. Furthermore, if a system starts out with an eccentric distribution (if all of the coins are heads up or if all of the air molecules are on one side) and random behavior takes place (the tray is shaken or the air molecules start to move), the distribution will shift toward an average distribution (nearly equal numbers of heads and tails or of molecules on each side). No amount of shaking (or random molecular motion) will tend to move the distribution back to an eccentric one. All of this is true only of isolated systems—if the system is not isolated, high-grade energy could be used to turn all of the coins heads up or pump all the air molecules to one side of the room.

In summary, the coin and gas molecule (Box 1.5) examples illustrate two important features of isolated, random systems containing large numbers of elements:

- Average behavior is much more probable than eccentric behavior.
- The direction of spontaneous change as a result of a random influence is from eccentric behavior (low probability) toward average behavior (high probability).

Another type of random behavior is the transfer of thermal energy between molecules as they collide with each other. If a large number of molecules in a room have a certain total fixed amount of thermal energy, there are many different ways in which that fixed amount of energy can be shared among the molecules. For example, one molecule might have all of it, each molecule might have a different amount, or the energy might be shared equally among them. Since the transfer of thermal energy during collisions is a random event, this situation is analogous to the coins on a tray and to the air molecules in a room. By analogy, then, there is an overwhelming bias toward average behavior (in this case, a roughly equal distribution of thermal energy among all of the molecules in a room), and eccentric behavior (one molecule possessing all of the energy, for example) effectively becomes impossible. Furthermore, the direction of spontaneous change is from eccentric behavior (an uneven distribution of energy) toward average behavior (an approximately equal distribution).

In this way, probability theory provides an explanation for the degradation of energy and also for the fact that energy degradation is irreversible. Friction (for example, air resistance) is the most common physical process responsible for the degradation—or dissipation—of energy.

Returning to our example of the fairground attraction, as the car pushes air molecules out of the way, a small amount of the its kinetic energy is transferred to the air molecules it collides with. Those molecules interact with other molecules, which interact with yet more molecules, and so on. If the fairground ride is outside, the situation is equivalent to a room containing all of the air molecules in the earth's atmosphere. Because of the higher probability of average behavior, the energy tends to be shared (dissipated) among an enormous number of molecules and, because the direction of change is from eccentric toward average behavior, the energy, once dissipated, cannot be recovered.

Other forms of friction have the same effect. As the car in the fairground attraction rolls along, it pushes on the rails. A small amount of kinetic energy is transferred to the iron atoms on the surface of the rails, and they heat up. Just like the air molecules, these iron atoms are in contact with other iron atoms in the rail and eventually the energy is dissipated throughout the rail.

In general, because the number of molecules in any macroscopic system is enormous, the laws of probability become laws of near certainty, and predictions based on those laws are entirely consistent with our observations. Therefore, the rules of the game effectively dictate that high-grade energy will eventually be dissipated into heat, and they prohibit the complete recovery of this low-grade form of energy.

Entropy

The second law of thermodynamics is often expressed in terms of the amount of "order" or "disorder" of a system, in which "order" is associated with low-probability (unlikely), regular arrangements of the elements (coins or gas molecules) that make up the system and "disorder" is associated with high-probability (likely), mixed-up arrangements of the elements. With this qualitative concept of order and disorder, the earlier statements concerning average and eccentric behavior of isolated, random systems containing large numbers of elements can be re-expressed as follows:

- Disorder is much more probable than order.
- The direction of spontaneous change as a result of a random influence is from order (low probability) toward disorder (high probability).

Since the universe is an isolated system, the second law asserts that the universe has a natural tendency to evolve in the direction of increasing disorder. This might be called the "homemaker's rule": order is improbable; disorder is probable. This is because there are many more ways to achieve disorder than there are to achieve order. Evidence of this tendency toward increasing disorder abounds in everyday life. Even without stirring, cold cream naturally mixes with hot coffee, eventually reaching a constant temperature and a uniform color. Hot coffee left to stand eventually cools to the air temperature of the room. A dropped egg naturally breaks open, spreading its contents and becoming more disordered. The reverse processes (the spontaneous reassembly of a smashed egg, the spontaneous heating up of a cold cup of coffee, or the spontaneous unmixing of cream and coffee) are impossible because they represent an increase in order of the system and, therefore, violate the second law. These reverse processes can be achieved, however, if the system is not isolated—for example, a chicken can lay a new egg, at the expense of increased disorder elsewhere.

In the study of thermodynamics, the concept of disorder is quantified. *Entropy* is a quantitative measure of the extent of disorder in a system—a measure of the number of ways a body composed of molecules in random motion can be mixed up inside and still look the same outside. A highly mixed up, disordered (high-probability) system has a high entropy, whereas a highly ordered (low-probability) system has relatively lower entropy. The fact that the total entropy of the universe is constantly increasing is another expression of the second law of thermodynamics. Entropy and energy both can be computed according to formulas, and the equations of thermodynamics show that entropy increases when heat (thermal energy) flows from a hot region or object to a cooler one. Because the increase in entropy is irreversible, heat can never flow spontaneously back from a cooler region to a hotter one.

Finally, the second law provides proof that perpetual-motion machines cannot be constructed. Since all machines in the real world involve some friction, the total amount of mechanical energy possessed by a machine must decrease and total entropy must increase when the machine is operated. It is rumored that Albert Einstein, while working as an examiner in the Swiss patent office in the early twentieth century, rapidly dismissed crackpot inventions as unworkable without looking at their details, because he was able to recognize very quickly that they relied on perpetual motion.

To summarize this section, the laws of probability show that the total amount of disorder in the universe increases irreversibly with every random event, because there are vastly more possible disordered arrangements for the universe than there are ordered ones. This is the same as saying that high-grade energy is dissipated irreversibly to a low-grade form in every energy transformation.

The Balance of Energy on Earth

The time scale for the dissipation of all of the high-grade energy in the universe is billions of years. Unfortunately, this is no consolation because the irreversible degradation of high-grade energy resources here on the earth is likely to have significant negative consequences much sooner—quite possibly within the next one hundred years.

Sources of high-grade energy from the earth, such as fossil fuels and uranium, are strictly limited. Most studies predict that the easily accessible sources of fossil fuels will be exhausted by the end of the twenty-first century, probably stimulating a permanent increase in the cost of energy. The dissipation of these high-grade energy resources through their use by humankind cannot be reversed (as dictated by the second law of thermodynamics).

However, there is some consolation to be found here because the second law of thermodynamics applies only to *isolated* systems and, fortunately, the earth is not an isolated system. In fact, the earth is constantly showered with an abundance of energy from the sun—2 trillion tons of burning nuclear fuel in the center of the sun, less than one hundred million miles away. Furthermore, the sun is expected to continue shining like this for another 5 billion years.

The problem of sustainability is, in fact, a rate problem. Our technology-based society has, over the past one to two hundred years, exponentially increased its consumption of high-grade energy. We have achieved this rate increase not by extracting more energy from our daily supply of sunlight but by exploiting energy stored in the earth as coal, oil, and natural gas by countless generations of plants and animals. Fossil fuel energy is stored solar energy. The storage process took hundreds of millions of years, whereas the extraction and burning of this stored energy by humans is occurring over a period of only a few hundred years.

Fortunately, our supply of energy from the sun still exceeds the current global demand for energy. But there are two major problems: (1) because it requires large collection areas (the equivalent of one-third of the state of

Wyoming to supply the current U.S. total energy needs), we are not currently equipped to utilize solar energy on a large scale, and (2) the global demand for energy is still increasing. To achieve energy sustainability, we must develop new technologies for utilizing solar energy in all its forms—direct solar radiation as well as wind and other forms of renewable energy that are driven by the sun. It is important to realize, however, that even renewables are not a panacea to our long-term energy problems. Solar energy, while widespread and long lasting, is dilute and limited in supply (see Chapter 2), and all alternative energy sources have environmental impacts of one kind or another (see Chapter 3). Long-term energy and environmental sustainability will require stabilization of energy consumption. Nuclear energy, which is uniquely exploitable by human beings, may have an increasingly important role to play in coming years during the transition from fossil fuels to renewables (see Chapter 2).

Exponential Growth

Continuing growth of world population and energy consumption per capita lies at the heart of world energy problems. The implications of steady growth—in particular, the enormous numbers that can be produced by seemingly small growth rates over modest periods of time—are not obvious or easy to grasp. Albert A. Bartlett has emphasized in his article "The Forgotten Fundamentals of the Energy Crisis" the urgent need for people to understand the arithmetic of steady growth.[7]

Most individuals, communities, businesses, investors, and governments regard steady growth (growth at constant percentage per year, also known as exponential growth) as a revered, almost sacred goal. As illustrated in Box 1.6, a steady growth rate in the value of an investment is certainly a good outcome for an investor. At an annual interest rate of 10 percent, the value of the investment doubles every seven years and increases a thousandfold over seventy years. (Note that exponential growth has a "doubling time" approximately equal to seventy divided by the annual percentage growth rate. This is known as the "Rule of 70.")

The large-scale consequences of steady growth of an undesired commodity can be, in the long run, quite alarming. Moreover, some alarming consequences can arise quite suddenly, as if without warning (see Box 1.7).

An economic growth rate of a few percentage points per year is the goal of many communities, businesses, and nations, and a small percentage growth rate like 2 percent sounds harmless enough. However, economic growth rates historically have correlated closely with growth in population, energy and natural resource consumption, and environmental pollution. That a steady exponential

Box 1.6. The "Magic" of Exponential Growth

If $1,000 is invested at a compound interest rate of 10 percent per year, the investment grows quite slowly at first. The investment only earns $100 in the first year, $110 in the second year, and $121 in the third year.

Year	Initial Balance	Interest Added	Final Balance
1	$1,000.00	0.1 x $1,000.00 = $100.00	$1,100.00
2	$1,100.00	0.1 x $1,100.00 = $110.00	$1,210.00
3	$1,210.00	0.1 x $1,210.00 = $121.00	$1,331.00
4	$1,331.00	0.1 x $1,331.00 = $133.10	$1,464.10
5	$1,464.10	0.1 x $1,464.10 = $146.41	$1,610.51
6	$1,610.51	0.1 x $1,610.51 = $161.05	$1,771.56
7	$1,771.56	0.1 x $1,771.56 = $177.16	$1,948.72

However, the interest added each year grows, so the principal on which the interest is calculated grows by an increasing amount each year. Because the basis for the interest calculation is increasing, compound interest (or exponential growth) can produce quite surprising results. For example, the initial $1,000 almost doubles in value in just seven years, at 10 percent interest. In general, compound interest always has a "doubling time" approximately equal to seventy divided by the annual percentage interest rate.

Now the investment really takes off. Since its value doubles in the first seven years, it must double again in the next seven years, and so on.

No. of Years	Final Balance
0	$1,000
7	$2,000
14	$4,000
21	$8,000
28	$16,000
35	$32,000
42	$64,000
49	$128,000
56	$256,000
63	$512,000
70	$1,024,000

Over seventy years, the investment doubles ten times, corresponding to a one thousand–fold increase in value, so the initial $1,000 investment is now worth over $1,000,000. It is interesting to note that the balance increases by twice as much in each seven-year period as it did in the previous one, and by more in the final seven years ($512,000) than in the entire preceding sixty-three years ($511,000).

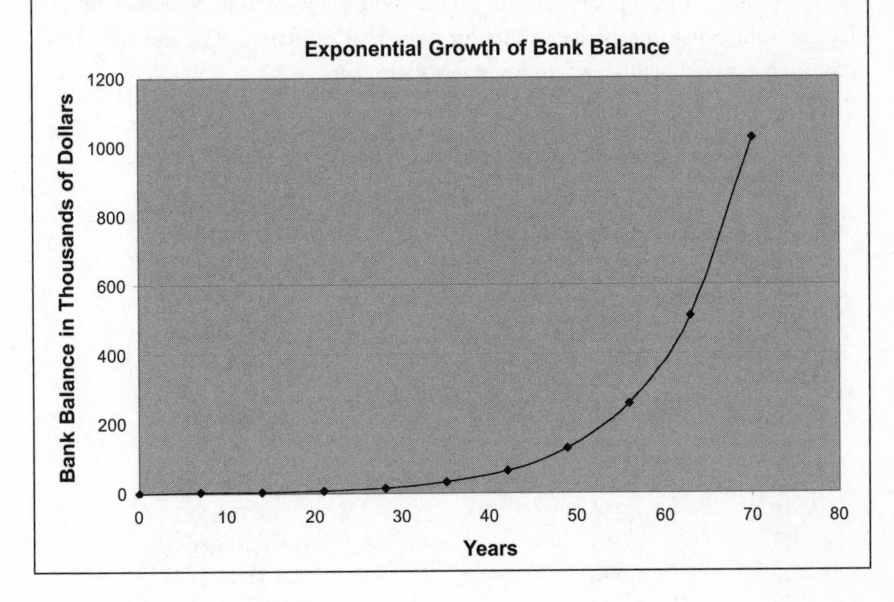

growth of 2 percent per year in any of these physical quantities is unsustainable is illustrated by the following example.

In 1975, the world population was about 4 billion people and was growing about 2 percent per year. This seemingly low annual growth rate of 2 percent corresponds to a doubling time of thirty-five years, implying that the world population would reach 8 billion by 2010. With a value of 6.5 billion in 2001 (as of this writing), the prediction is right on. If this growth rate should continue, the world population would double again (to 16 billion) by the year 2045—and so on. Within five hundred years, the population would exceed 200 trillion and there would be, on average, one person per square meter of dry land surface! Small growth rates can yield incredible numbers in modest periods of time.[8]

Clearly, steady population growth—even at a small annual percentage rate—is not sustainable for long on a finite planet. Another alarming characteristic of exponential growth is that it can easily go unnoticed until the last few doublings. As shown in Box 1.7, lily pads that double in population every year might go unnoticed for more than twenty-five years before suddenly filling a lake in just a few more years.

Historically, economic growth and resource consumption have been closely coupled. The danger of steady exponential (constant percentage) growth, which

Box 1.7. The "Hidden Danger" of Exponential Growth

Lily pads live on a certain lake which is so nurturing that the total number of lily pads doubles each year. If two lily pads live on the lake in the first year, four in the second year, eight in the third year . . . etc., the number reaches 2^{30} (about one billion) after thirty years and let us suppose the lake is now completely covered. Note how the lake changes from being mostly open water for more than a quarter of a century to being completely covered by lily pads in just a few more years.

EXPONENTIAL GROWTH OF LILY PADS ON A LAKE

Year	No. of Lily Pads	Year	No. of Lily Pads
1	2	16	65,536
2	4	17	131,072
3	8	18	262,144
4	16	19	524,288
5	32	20	1,048,576
6	64	21	2,097,152
7	128	22	4,194,304
8	256	23	8,388,608
9	512	24	16,777,216
10	1,024	25	33,554,432
11	2,048	26	67,108,864
12	4,096	27	134,217,728
13	8,192	28	268,435,456
14	16,384	29	536,870,912
15	32,768	30	1,073,741,824

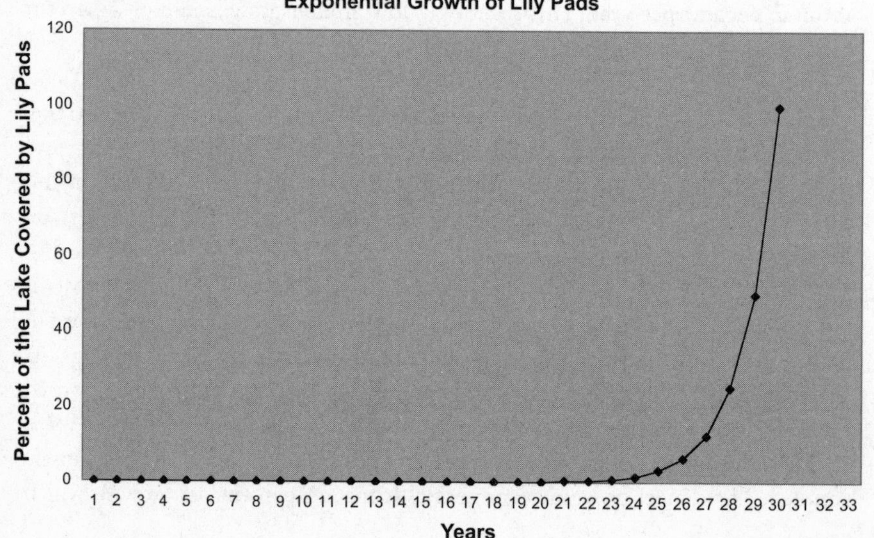

the lily pads example illustrates, is that we may not see the problems coming until late in the game, when the time left to adjust is very limited. The keys to adjustment will be population control and breaking the connection between economic growth and growth in resource consumption and environmental pollution. This issue is discussed in detail in Chapter 6.

Exponential growth is a realistic model of growth when resources are abundant, but steady growth of this type that is dependent on resource consumption is not sustainable indefinitely in a finite environment. Under real-life market conditions, exponential growth does not continue to the point of complete exhaustion of a resource, but the sudden onset of the limitations to growth characteristic of exponential growth can cause immense suffering. It is important to understand the nature of exponential growth (in particular the "Rule of 70," which is illustrated in Box 1.6) in order to anticipate and make adjustments to the limits of this kind of growth.

Conclusion

In this chapter we focused on the key scientific ideas that are central to dealing with energy problems. Most subtle and least well understood are the second law of thermodynamics, explaining that useful energy is a limited commodity that is depleted in all interactions, and exponential growth, which is never sustainable and can lead to unexpected and critical resource shortages. Policy makers and the general public must understand these two ideas, as they are vital to developing sound energy policy and ways of life consistent with long-term energy sustainability.

Key Ideas in Chapter 1

- The laws of nature are absolute. They cannot be circumvented by human cleverness or technology.
- The two essential laws regarding energy are the first and second laws of thermodynamics.
  - According to the first law of thermodynamics, energy cannot be created or destroyed—the energy of the universe is constant.
  - According to the second law of thermodynamics, the conversion of high-grade energy to low-grade energy is irreversible and unavoidable—the entropy of the universe tends toward a maximum.
- Energy comes in many forms.

- Fossil fuels and nuclear energy sources are limited and nonrenewable (on a human time scale). As they become depleted, humans will be forced to learn how to live on other forms of energy.
- Solar energy (direct radiation as well as wind and other forms of renewable energy that are driven by the sun) is widespread and will last as long as the sun continues to shine (billions of years). However, meeting large-scale energy needs with solar energy requires large collection areas because of its low concentration.
- Steady growth in world energy use, as for the use of any physical resource, is not sustainable.
- A "magic bullet" in the form of unknown energy sources and technologies that could permanently resolve world energy problems is unlikely.

Notes

1. Richard P. Feynman, *Six Easy Pieces* (Reading, Mass.: Addison-Wesley 1995), ch. 4; Richard P. Feynman, Robert B. Leighton, and Matthew Sands, *The Feynman Lectures on Physics* (Reading, Mass.: Addison-Wesley, 1963), ch. 4.
2. Although nonrenewable, deuterium could provide a virtually limitless supply of energy if a process to extract and utilize it could be developed (see Chapter 2).
3. Robert Resnick and David Halliday, *Physics* (New York: Wiley & Sons, 1977); Paul A. Tipler, *Physics for Scientists and Engineers* (New York: Worth, 1991); Jack J. Kraushaar and Robert A. Ristinen, *Energy and Problems of a Technical Society* (New York: Wiley & Sons, 1988); Joseph Priest, *Energy: Principles, Problems, Alternatives* (Reading, Mass.: Addison-Wesley, 1991).
4. Morton Mott-Smith, *Principles of Mechanics Simply Explained* (New York: Dover, 1963); Morton Mott-Smith, *The Concept of Heat and Its Workings Simply Explained* (New York: Dover, 1962); Morton Mott-Smith, *The Concept of Energy Simply Explained* (New York: Dover, 1964).
5. Resnick and Halliday, *Physics.*
6. See Kenneth W. Ford, *Basic Physics* (Waltham, Mass.: Blaisdell, 1968), ch. 14.
7. Albert A. Bartlett, "Forgotten Fundamentals of the Energy Crisis," *American Journal of Physics* 46 (1978): 876–888.
8. Bartlett, "Forgotten Fundamentals of the Energy Crisis."

Chapter 2

Future World Energy Needs and Resources

JOHN SHEFFIELD

We have established the centrality of energy sources and energy flows in sustaining all life and life-related activity. This is a fundamental point because energy flows are ultimately finite and move only in one direction—from the useful to the useless. These flows are subject to laws of nature that cannot be altered by human intervention. We are thus subject to some limits in our ability to provide adequate energy to meet human needs and aspirations for a satisfactory worldwide standard of living that can be sustained beyond the foreseeable future. This is the anchor—the reality check for all projections and policy prescriptions.

Most fundamentally, these rules relate to the growing number of people on earth and their growing appetite for transforming energy to meet human needs and wants. The resulting demand for energy must be set against the energy resources and the present and anticipated future technology available to transform and "use" this energy.

The most developed countries of the world use a heavily disproportionate share of energy—especially on a per capita basis. The greatest increases in world energy use in coming years, however, will be in developing countries because of the rapid rates of growth of population and consumption in these countries. Will the world's finite energy resources and new technology options be adequate to meet the resulting increase in demand for energy?

In this chapter, John Sheffield addresses the problem of meeting future energy needs in a world in which both population and energy use continue to grow. We have some useful historical knowledge of the energy required to increase standards of living to levels that seem to be compatible with stabilizing population size. This in turn gives us insight into the energy requirements to meet our sustainability goal.

—Editors' note

Table 2.1. World energy use in 1996

Population group	Population (billions)	Total energy use (Mtoe/a)	Avg. energy use per capita (toe/cap·a)
World, total	5.7	9,600	1.7
Developed nations	1.3	6,200	5.0
Developing nations	4.4	3,400	0.8 [a]

[a] For commercial energy alone, an average of only 0.6 toe/ cap·a.

Population growth and, in particular, the consequences of an increased demand for energy to meet the aspirations of the developing world are the fundamental human factors in understanding energy.[1] In late 1996, the world population of some 5.7 billion people used about 8,500 million tonnes of oil equivalent of "commercial" energy per year (Mtoe/a) plus about 1,100 Mtoe/a of biomass energy (agricultural and forest residues, manure, and a few energy crops sometimes described as "noncommercial"). (See Appendix 2 for definitions of Mtoe and other energy units used in this chapter.) This was an average of 1.7 toe per capita per year (toe/cap·a) in total (see Table 2.1). The most striking feature of this energy use is that while the developed world's 1.3 billion people used about 6,200 Mtoe/a—in other words, about 5 toe/cap·a—the developing world's 4.4 billion people used about 3,400 Mtoe/a, for an average of only 0.6 toe/cap·a of "commercial" energy and a total of about 0.8 toe/cap·a, including biomass energy, which is often used ineffectively.

In *Energy as an Instrument for Socio-Economic Development,* Goldemberg and Johansson suggest that the 1 toe/cap·a energy use level is an important threshold for societal development.[2] They point out that while low energy consumption is not the cause of poverty, it is an indicator for many of its elements, including poor education, bad health care, and the hardship imposed on women and children. The significance of annual per capita energy use lies in its apparent relationship to fertility rates in women, as discussed below.

Goldemberg and associates estimate that basic human needs might be met in a warm climate country with as little as 0.8 toe/cap·a, *provided the energy were used efficiently.*[3] Allowing that not everyone in the developing world lives in a warm climate and that more than the bare minimum should be the goal, a minimum energy goal for a decent standard of living might be 1.5 toe/cap·a in the developing world. If the efficiency of energy use were doubled, the minimum energy standard of 1.5 toe/cap·a of the future would be equivalent, in useful energy, to 3 toe/cap·a today, which is at the low end of energy use in many developed countries. See Table 2.1 for a comparison of current total energy use and energy use per capita in developed and developing nations.

Population Growth and Per Capita Energy Use

The population of the world, if we discount the brief decline caused by the Black Death, has increased steadily since at least the time of the Roman Empire. Today, the world has about 6 billion people, and there is no agreement as to when the population might stabilize and, if it did stabilize, at what level that would be. The World Bank and the United Nations estimate, on the basis of studying population trends in every country, that world population might stabilize at around 9 to 12 billion people early in the twenty-second century, depending on the rate of decline of the fertility rate (the expected number of births for a woman surviving to the end of the childbearing period), as indicated in Figure 2.1.[4] Other estimates vary from an optimistic view that the population will remain around today's level to projections of 20 billion or more people.

The current growth rate of world population is about 1.5 percent per year, with signs of a small rate of decrease, a trend in line with the middle-of-the-road

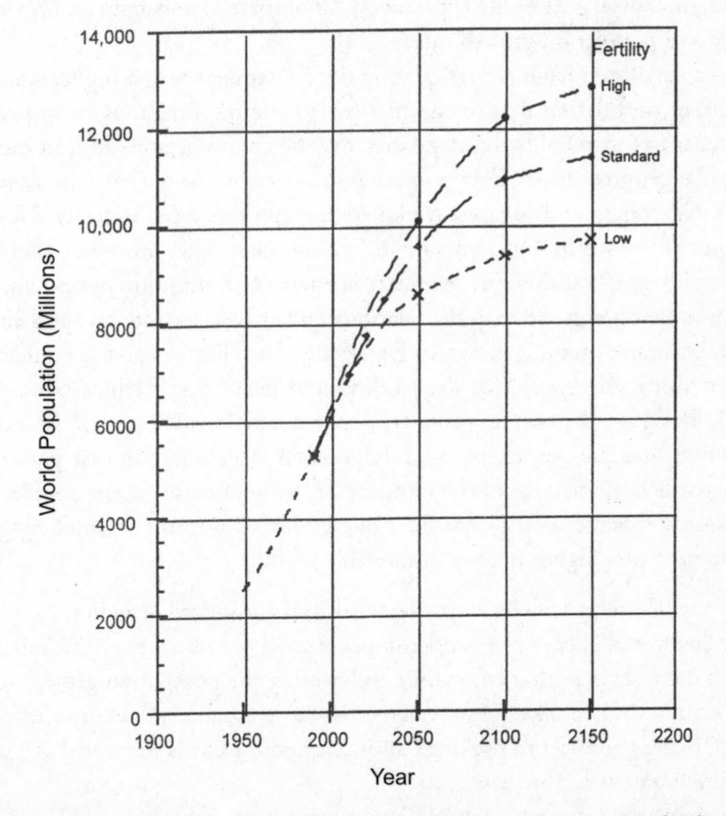

Figure 2.1. World Bank population estimates to 2150, based on varying fertility rates. *Source:* E. Bos et al., *World Population Projections: 1994-95 Edition.*

estimates. Certainly, in the developed world the population is more or less stable at about 1.3 billion people. Some of the more affluent countries, such as Italy and Spain, and countries undergoing traumatic transitions, such as Russia, have an average birth rate even lower than the replacement rate of about two children per woman. In the developing countries, growth rates are high—as much as 3 to 4 percent per year. By the Rule of 70 (doubling time in years $\approx$70/annual percentage growth), this growth rate, if sustained, will mean a doubling of the population in developing nations every seventeen to twenty-three years.

As Cohen notes in his book *How Many People Can the Earth Support?* there is no demonstrated, unambiguous connection between fertility rate and standard of living.[5] However, with an increasing standard of living come more extensive education for both men and women, increasing literacy, more rapid availability of information (by print, radio, or television), greater availability and use of contraception, more opportunities for women, lower infant mortality, less need for parents to want a large number of children to support them in their old age and, of course, greater life expectancy. All of these factors have an impact on fertility and population growth rates.

It is a small step from contemplating the consequences of a higher standard of living to postulating that the annual energy use per capita, as an important determinant of material living standards, may be an important factor in the rate of population growth. In *Energy as an Instrument for Socio-Economic Development,* Goldemberg and Johansson plot on a graph the social indicators for the countries of the world—literacy rate, infant mortality rate, life expectancy, and total fertility rate—against the commercial energy consumption per capita. The portion of that graph showing the relationship between fertility rate and energy consumption per capita is shown in Figure 2.2. The chart suggests a connection between rising energy use (or availability) and decreasing fertility rates. Note that in this figure, the outlying countries are a relatively small part of the world's population and are, generally, oil producers in which the annual per capita energy use is not reflective of the standard of living of many of the people.

It seems sensible then to consider energy use as a factor in discussing the development of a higher-quality, sustainable world:

- The use of energy enables improvements in the standard of living.
- The absence of energy can condemn people to a poor quality of life and deny them those services that contribute to lowering the population growth rate.
- An energy unit, unlike a monetary unit, has the same value anywhere it is used, making it easier to use in a consistent fashion across the world as a state-of-development factor than money.
- If energy use influences population growth rate, there can be significant impacts from variations in its availability and deployment.

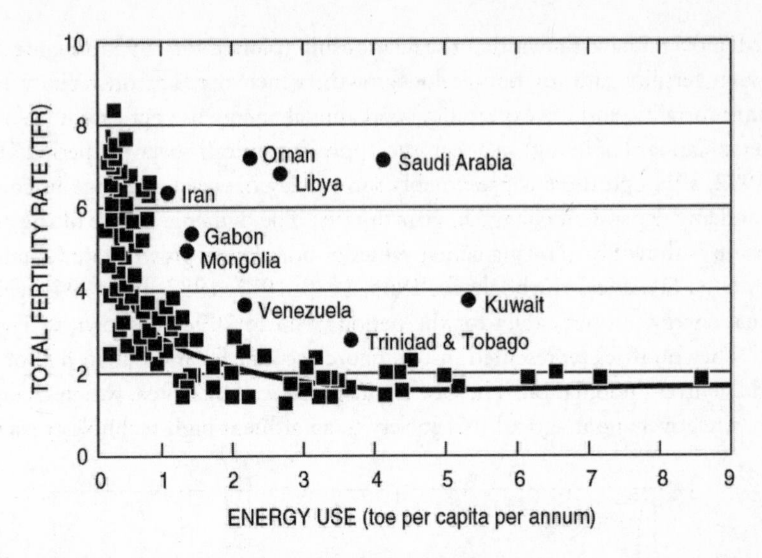

Figure 2.2. Fertility rate as a function of commercial energy consumption per capita. *Note:* Noncommercial fuels are not included. *Source:* Goldemberg and Johansson, *Energy as an Instrument for Socio-Economic Development.* Reproduced courtesy of Goldemberg.

For these reasons, Holdren, Gouse et al., and Sheffield consider annual per capita energy use in projecting future world energy needs.[6]

Energy availability and use, however, is only one factor affecting human well-being and economic development. Culture is another important factor that determines behavior and, therefore, development. Duchin uses two similar factors, technology and lifestyle, as organizing concepts for understanding the processes in global development.[7] Technologies are important because the effective production and use of energy requires the development and deployment of improved technologies. In addition, the price of energy and energy resources is a factor in whether a particular future will be realized in each part of the world if population and energy demand increase as they are expected to.

Thus, annual energy use per capita is an interesting surrogate factor that may capture the standard-of-living aspects of population growth in each country or area of the world. If per capita energy use does serve this function, then the current usage figures, discussed above, indicate a problem for developing areas of the world. Stabilization of the population in these regions will require more energy per capita *now.* Failure to raise energy use per capita can lead to sustained high rates of population growth and to an even higher energy use later. It is not clear whether cultural changes can work rapidly enough to lower the rate and stabilize population without the accompanying improvements in standard of living bound up with increased energy use.

Moreover, I have shown that the relationship (plotted for 1993 in Figure 2.2) between fertility rate (or population growth, which depends on fertility rate, infant mortality, and life expectancy) and annual energy use per capita (a surrogate for standard of living) also seems to apply dynamically over the period 1965 to 1992, although there is presumably some delay between a change in energy use and any dependent change in growth rate.[8] The dynamic nature of this relationship is shown by plotting actual values of population growth rate (obtained from U.N. Statistical Yearbooks for 1965, 1975, 1977, 1987, and 1994) against annual energy use per capita for the period 1965 to 1992, as shown in Figure 2.3.[9] The countries represented in this figure account for more than half of the world's current population. The case is illustrated well by Korea, which changed from a relatively poor agricultural society to an affluent high-technology society

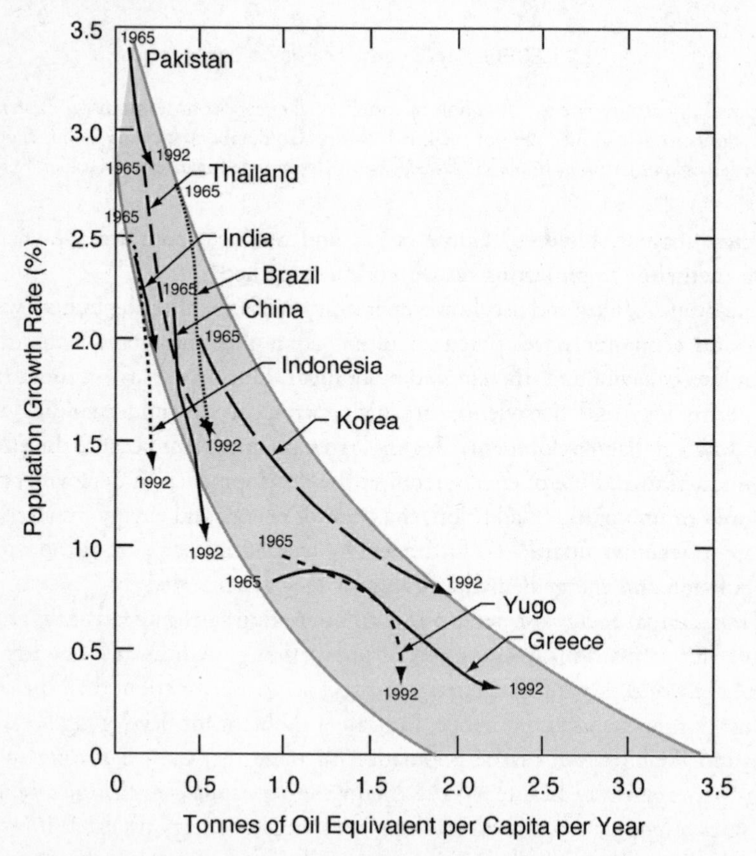

Figure 2.3. Actual population growth rate plotted against total energy use per capita for selected countries, 1965–92. *Source:* data obtained from United Nations, *Statistical Yearbook,* 1965, 1975, 1977, 1987, 1994.

during this period. For Latin America, Brazil, with the largest population, shows the general trend. Other Latin American countries have not changed much over this period, either in per capita energy use or growth rate, but are close to the Brazilian curve shown in the figure.

Inspection of Figure 2.3 shows differences in growth rate for a given per capita energy use that may be viewed as cultural. Is it possible that the countries with a relatively higher growth rate will achieve a "cultural" factor comparable to that of India and China today toward the end of the next century? Certainly, in regard to cultural gains, one has to look no farther than the variations in growth rate between different countries, or even within a country, to realize the potential for improvement. Kerala State in India is often cited as an example of how much can be done to deliver a high quality of life with a low gross national product per capita (though in this case some allowance must be made for funds obtained by citizens working abroad).[10] There is also evidence for a decline in fertility in most developing countries.[11]

Africa is not represented in Figure 2.3 because, to date, there has been only a small move to a decreasing population growth rate, despite the ravages of AIDS. In Africa, the population growth rate increased slightly from the 1950s into the 1970s and held at around 3 percent per year into the 1980s.[12] The increase in growth rate was mainly because of declining mortality due to better health care. Only recently do most of the African countries show a downturn in population growth rate. So, in the absence of a long-term decline in population growth, the question remains as to where Africa might lie on this chart if its per capita energy use continues to increase.

World Energy Demand Projections

In the standard fertility case estimates of the World Bank and the United Nations, the developing world's population alone will increase from 4.7 billion to about 9.6 billion people by 2100. This increasing population will make increasing demands on resources, including energy. Thus, key questions in regard to future energy needs include the following:

- Is there a sustainable way to provide the energy resources needed for a decent standard of living for everybody in a world population that is nearly double today's population?
- If there is a connection between a decreasing population growth rate and increasing energy per capita, how much energy do the developing countries need to match the projections of growth rate decline?
- What are the energy resources, and how do the resources in each area of the world match the needs of that area?

Table 2.2. Some projections of energy demand (Mtoe/a)

Case	Year 2010	2050	2100
Low	12,000	14,000	16,000
Medium		18,500	21,000
High		22,500	37,500

Sources: Holdren, "Energy in Transition"; International Energy Agency, *World Energy Outlook*; International Panel on Climate Change, *Climate Change, 1995*; Gouse et al., "Potential World Development"; Sheffield, "Population Growth"; International Institute for Applied Systems Analysis and World Energy Council, *Global Energy Perspectives to 2050 and Beyond* (London: World Energy Council, 1995).

- Is there something special about the nearer-term situation, in which there is a flood of cheap oil and gas?
- How can the use of a wider range of energy sources and a more efficient use of energy help meet the needs?

Projections for total energy use by the end of the present century are generally in the range of 20,000 to 45,000 Mtoe/a. Some reference energy projections are listed in Table 2.2 and are plotted schematically in Figure 2.4.

Table 2.3 shows world energy use in 1996 by energy category for the Organization for Economic Cooperation and Development (OECD) countries (OECD North America, including Mexico; OECD Europe; and Pacific OECD) plus the Former Soviet Union (FSU) and Central and Eastern Europe (CEE)—roughly the developed regions—compared with that for the rest of the world. As can be seen, the total world energy use for that year was approximately 9,600 Mtoe.

The energy supply for the developing world alone would have to rise to 48,000 Mtoe/a for each person to have 5 toe/cap·a in the standard population case of the World Bank—more than five times today's total energy use. As is discussed in the following section, such a level of energy use might be achieved for a short period of time using a wide variety of sustainable and nonsustainable energy sources. But it does not appear to be sustainable, short of a massive availability of solar and nuclear (fission and fusion) energy. A more likely solution to providing the energy needed to give everybody a decent standard of living is through much-improved energy efficiency. It seems reasonable to assume that what matters to providing a good standard of living is the useful energy, not the wasted energy. If this is the case, an average improvement in the efficiency of energy use by a factor of 2 would mean that 48,000 Mtoe/a with today's

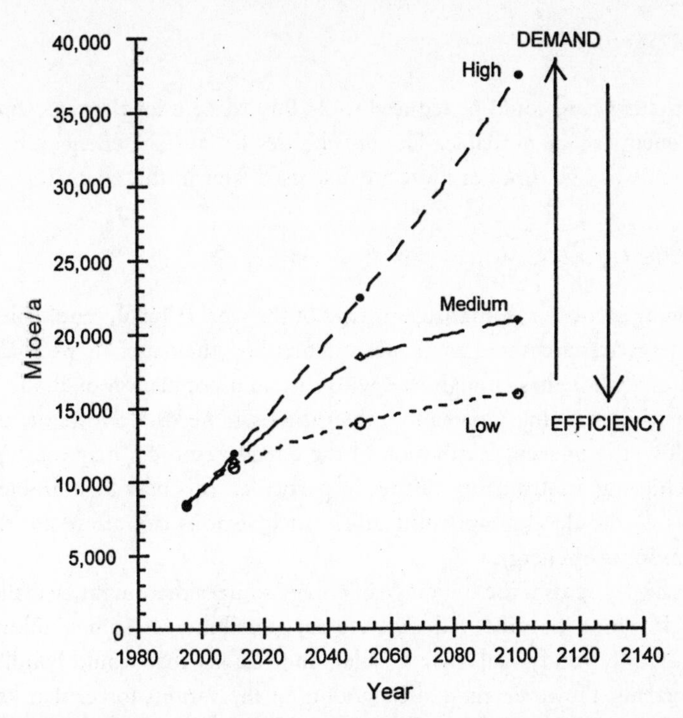

Figure 2.4. Projections of energy demand for the twenty-first century. *Sources:* Gouse et al. (n. 6), Holdren (n. 6), Sheffield (n. 6), IPCC (n. 17), IASA/WEC, *Global Energy Perspectives to 2050 and Beyond,* (London: World Energy Council, 1995). (See Table 2.2)

Table 2.3. Approximate energy use (Mtoe) by energy category for developed and developing countries, 1996

Energy source	Developed countries (OECD + FSU + CEE)	Developing countries	World total
Coal	1,430	1,030	2,460
Oil	2,320	1,010	3,330
Gas	1,530	330	1,860
Nuclear	560	30	590
Hydropower	130	90	220
Geothermal, solar, tidal, etc.	30	10	40
Biomass crops and wastes	200	900	1,100
All sources	6,200	3,400	9,600

Sources: A composite of data and projections from IEA 1995 and 1998. *World Energy Outlook, 1995 and 1998 Editions.* Paris: OECD Publications. WEC, *1995 and 1998 Survey of World Energy Resources,* London: World Energy Council. Johansson, T. B., H. Kelly, A. K. N. Reddy, and R. H. Williams, eds. 1993. *Renewable Energy: Sources for Fuels and Electricity.* Washington, D.C.: Island Press. Larson, E. D., and R. H. Williams. 1995. "Biomass Plantation Energy Systems and Sustainable Development." In J. Goldemberg and T. B. Johanson, eds., *Energy as an Instrument for Socio-Economic Development.* New York: U.N. Development Programme.

inefficient energy use could be reduced to 24,000 Mtoe/a for the same amount of useful energy to be available. The possibilities for average energy efficiency improvements of two times or more are discussed later in this chapter.

World Energy Resources

The following sections summarize estimates of the world's fossil, renewable, and nuclear energy resources and the issues surrounding their use. As we shall see, the world as a whole has enough energy to sustain a population of about 10 to 12 billion while meeting rather more than the basic needs.[13] However, as discussed below, the uneven distribution of the energy resources may cause problems in achieving a satisfactory future. In particular, this may be a problem in some parts of the developing world unless such regions can afford to import massive amounts of energy.

Appendix 1 discusses the full range of energy sources that might be exploited on earth. It would be nice if there were the possibility of some "unknown" energy source (beyond fossil, fission, solar, and fusion) that would handle any future shortfalls. However, the understanding of the various forces that lead to energy is very well advanced and the fusion of light elements appears to be the only untapped energy source. In fact, both solar energy—including its manifestations as biomass, wave, wind, and direct heat and light—and the fusion of deuterium are effectively resources for all time—or at least until the sun expands and envelops the earth. (See Table 2.4 for a summary of world energy resources.)

Environmental considerations are crucially important in the production and

Table 2.4. Potential world energy resources in millions of tonnes of oil equivalent per year (Mtoe/a)

Energy type/resource	Amount of energy[a] (Mtoe/a)	Comments	Source(s) of estimate
FOSSIL ENERGY[b]			
Oil	3,000 (t)	Duration: 100 yrs.	Grubler, Jefferson, and Nakicenovic, "Global Energy Perspectives"
Gas	4,200 (t)	Duration: 100 yrs. for conventional gas	
Coal	3,400 (t)	Duration: 1,000 yrs.	
Shale oil + bitumen	1,100 (t)	Duration: 100 yrs. at 20% recovery	
Subtotal	11,700 (t)		

Energy type/resource	Amount of energy[a] (Mtoe/a)	Comments	Source(s) of estimate
RENEWABLE ENERGY			
Biomass	4,500 (t)	Issue of balancing tradeoff against food production[c]	Hall et al., "Biomass for Energy", Larson and Williams, "Biomass Plantation"
Geothermal	~300 (t)	500 x 10^{18} J of resources, replenished over 40 years	Palmerini, "Geothermal Energy"
Hydropower	800 (e)		WEC 1995, 1998
Solar	4,000 (e)	Assuming collectors on 0.1% of land area generating an average of 40 W(e)/m^2	
Wave + tide	>100 (e)	Estimate makes allowance for difficulties in realizing full potential of this energy source	World Energy Council, *Survey of Energy Resources,* 1995, 1998
Wind	4,500 (e)		Grubb and Meyer, "Wind Resources", Sorensen, "History"
Subtotal	5,000 (t) 9,400 (e)		
NUCLEAR ENERGY			
Fission (uranium breeder reactor)	3,000 (e)	Duration: 1,000 yrs. at <$130/kg (without thorium and uranium from seawater)[d]	Grubler, Jefferson, and Nakicenovic, "Global Energy Perspectives", World Energy Council, *Survey of Energy Resources,* 1995, 1998
Fusion:			
Deuterium-tritium		Limited by lithium supply	
Deuterium cycle		Unlimited	
Total energy	16,700 (t) 12,400 (e)		

[a] t = thermal energy; e = electric energy.

In comparing future energy use with today's energy use of about 9,000 Mtoe/a, the electrical use may be interpreted in different ways. As a replacement for, or complement to, electricity generated from fossil fuel (presumed to be produced at about 50% efficiency in the future), it may be viewed as worth about twice the electrical value in terms of thermal energy. As a source of hydrogen to replace fossil fuels (from electrolysis, for example), it would be worth less in thermal energy.

[b] Much larger amounts of fossil energy exist than are tabulated here, but these sources have a more speculative recovery potential; see the discussion on pages 2–9 ff.

[c] There is some debate about how much biomass energy might be available, even with gains in crop productivity, when allowance is made for food needs and degradation of arable land (Pimentel, "Biomass Fuel;" "Environmental and Economic Costs"). However, as much as half of the energy shown could be available from agricultural and forestry residues (Hall et al., "Biomass for Energy).

[d] If a major breeder reactor system is deployed, the cost of electricity should not be very dependent on the uranium cost; therefore, it would be economic to use the much larger but more costly land-based uranium resource of about 11,000,000 tonnes. Further, in principle, it would be possible to extract vast quantities of uranium from seawater if the uranium were used in breeder reactors.

use of the various energy sources, as discussed by Russell Lee in Chapter 3. All energy sources have an impact on the environment through their varied use of land, mining, transportation, component manufacturing, and solid, liquid, and gaseous wastes. An optimized deployment of energy sources will include a careful analysis of the tradeoffs among availability, energy security, cost, and environmental impacts.

Fossil Energy

When discussing fossil fuels, we tend to think in terms of conventional oil and gas. These resources are actually relatively small compared to today's annual use—less than one hundred years' worth. However, total estimated fossil resources including coal are possibly thirty times bigger (see Table 2.4). Thus, an issue for the future is not that fossil fuels will run out in the next century but, rather, that they may become more expensive as easy-to-obtain resources are depleted. This increasing cost may widen the gulf between those who have fossil resources and those who do not, and also between the rich and the poor.

However, improvements in the technologies of extraction, conversion, and end use will increase the resource base and allow more of the resources to be economical. Such improvements include enhanced oil discovery and recovery techniques, more efficient gas-to-liquid conversion, high-efficiency turbines and fuel cells, and lower-cost pollution removal systems. The choice, then, to restrict or sustain high fossil fuel use presumably will depend on the response to pollution and global warming concerns. In this regard, if a low or moderate cost of removal and safe sequestration of carbon dioxide can be realized, global warming concerns will not set much of a limitation on fossil fuel use for the next century or so.

On the basis of World Energy Council estimates, Grubler, Jefferson, and Nakicenovic assess the recoverable resources of conventional oil and natural gas liquids as about 295,000 Mtoe and of natural gas resources as about 420,000 Mtoe.[14] Shale oil and bitumen amount to another 560,000 Mtoe, of which some 20 percent or more might be economically recoverable. If the present rate of use of oil and gas (about 5,500 Mtoe/a) were continued, these resources would be depleted around 2150. In addition, as smaller oil fields are used up, resources increasingly will be concentrated in only a few places, notably the Middle East. In reality, unless global climate change considerations limit fossil fuel use, the rate of use will continue to increase. With a 2 percent per year increase, these resources would be depleted around 2070.

However, there would still be substantial coal reserves in 2100. These reserves are estimated at about 3,400,000 Mtoe today, with further, speculative resources of 3,000,000 Mtoe of coal. Therefore, a possible solution to energy needs over the next century would be the greater use of coal. In addition, there exist other, more speculative fossil fuel resources, including 1,900,00 Mtoe of oil, 850,000 Mtoe of gas, and up to 18,000,000 Mtoe of methane hydrates in tundra regions and the sea.[15] In addition, carbon dioxide injection into unmineable coal seams may be used to release trapped methane at the same time that the carbon dioxide is being sequestered.

The huge methane hydrate deposits are of particular interest because methane produces the least carbon dioxide of the fossil fuels and because the deposits are widespread off the coasts of a number of countries. For example, in the United States, large deposits exist off the coast of North Carolina, in the Blake Ridge. Estimated U.S. deposits are about one hundred times that of conventional natural gas. Key issues for exploiting this resource are economics and safety, and more research and development will be needed to determine how much can be safely mined and at what cost.

GREENHOUSE GAS ISSUES

During at least the past one hundred years, the average temperature of the atmosphere at the earth's surface has been increasing. Debate continues as to whether this increase is part of a natural cycle or is caused by human-made greenhouse gas emissions (notably carbon dioxide and methane) mainly from the use of fossil fuels. Either way, it seems sensible to be prepared to reduce emissions if the evidence that the temperature rise is human induced becomes stronger. Further warming is expected to lead to significant changes in the weather patterns, which affect agriculture, and to rises in sea levels, which lead to problems for the many people who live in low-lying coastal areas. As can be seen in Table 2.3, fossil fuels provide most of today's energy, around 8,000 Mtoe/a.

Carbon sequestration—the use of techniques that capture and provide long-term storage of carbon gases, such as carbon dioxide, to remove them from the atmosphere—is a very important consideration if global warming concerns are taken seriously. Without sequestration, fossil fuel use might have to be reduced by the end of the next century to 3,000 to 4,000 Mtoe/a.[16] This topic has been extensively discussed by the International Panel on Climate Change (IPCC) that has been studying all aspects of climate change for a number of years, and making projections of future energy use and emissions.

If the present levels of fossil fuel use continue for the next century, carbon emissions will be around 700 GtC (gigatonnes of carbon) in the period to 2100, of which perhaps half would have to be sequestered to limit carbon dioxide concentrations to less than 450 ppm (parts per million). In terms of fossil fuel use, the example cases of energy use discussed below lie between the IPCC cases IS92c and IS92d.[17]

Carbon may be sequestered many ways, including burial in depleted oil and gas wells, deposition in *saline aquifers* (salty water regions below the earth's surface), and deposition in the oceans.[18] These three approaches are being used today. It is estimated that oil and gas wells could sequester 130 to 500 GtC, while saline aquifers might sequester 90 to more than 1,000 GtC; the oceans could potentially handle 400 to more than 1,200 GtC. Important practical questions of cost and safety remain to be resolved for the ocean and aquifer solutions before massive sequestration will be allowed.

Renewable Energies

Until the advent of the massive use of fossil fuels, renewable energies—such as biomass for cooking and heating, hydropower for driving grain mills, and wind energy for ships—were the mainstays of humanity. Geothermal energy was used for hot baths, and solar heating was used for drying crops. Hydropower has been an important part of the electrification of the world since the nineteenth century.

More recently, solar and wind-generated electric power have become more important, and tidal and wave power has been exploited. The potential energy available from solar, wind, and wave resources is constrained by the issue of storing their "intermittent" electric power. Much debate exists about how much intermittent electrical power could be incorporated into an electrical supply system without a storage system. Estimates range from 5 to 45 percent, with a typical value of 15 to 20 percent.[19]

The availability of renewable energies varies widely from region to region because of differences in geography, topography, and climate. For example, solar energy intensity is stronger near the equator than near the poles.

BIOMASS ENERGY

Biomass energy in various forms—wood, crops, grass, animal wastes—accounts for more than 10 percent of energy use today (more than 1,000 Mtoe/a). It is estimated that an intensive use of biomass energy could produce 4,500 Mtoe/a.[20] Of this energy, about half would come from agricultural residues, farm animal wastes, forestry residues, and landfills, with the rest from dedicated energy crops. In such biomass-intensive scenarios, care would have to be taken to satisfy agricultural needs and ensure sustainable land use.[21] This biomass may be burned to produce heat or may be digested or heated to produce such gaseous fuels as methane, hydrogen, and carbon monoxide. It may also be converted to liquid fuels, such as ethanol, methanol, or biodiesel. Although using biomass energy releases carbon dioxide, the new plant growth absorbs most of the emissions. Therefore, biomass energy is viewed as a low emitter of greenhouse gases.

Biomass residues are large where agriculture and forestry production is high. Biomass energy crops are viable mainly in lower population density areas where there is a surplus of fertile land—for example, Brazil and the United States.

GEOTHERMAL ENERGY

Geothermal energy is highly concentrated around the Pacific Rim; on a band from the Mediterranean through the Himalayas; in Eastern Africa, Central Asia, and numerous Pacific islands; and in Iceland. Palmerini estimates that the total geothermal resources close enough to the earth's surface to be exploited are about 500×10^{18} J (joule) of energy.[22] If this were replenished every forty years, it would provide about 300 Mtoe/a. Geothermal energy has important niche possibilities in many countries and is presently providing about 10 GW(e) (gigawatts of electricity) and about three times as much heat energy.[23] As a source of energy to heat pumps, it has a much wider range of applicability.

HYDROPOWER

Large hydropower resources require high levels of rainfall and a substantial vertical drop between the source and the sea. The economically exploitable potential is particularly high in Brazil, Canada, China, India, Peru, the Russian Federation, the United States, and Zaire. World hydropower use is expected to increase. About 90 percent of the potential resources estimated by the World Energy Council are listed in Table 2.4.[24] The concerns related to hydropower use—damage to land, methane emissions from rotting vegetation that becomes submerged, and eventual silting—may limit the amount of the resource that is used. But for most areas of the developing world, hydropower is an important resource.

SOLAR POWER

Outside the earth's atmosphere, *insolation* (the sunlight energy per unit area) is a maximum of 1.37 kilowatts per square meter (kW/m^2), which atmospheric absorption reduces to a peak of about 1 kW/m^2 at the earth's surface, on a clear day, at midday and normal incidence. The daily average is less, and at higher latitudes, insolation decreases. The highest yearly average insolation levels in the U.S. Southwest are about 320 watts per square meter (W/m^2), with levels decreasing to about 140 W/m^2 around New York.

Numerous options exist for capturing solar energy: heating water for buildings; setting up a modest temperature gradient in a pond of salt water and using

Figure 2.5. Photograph of photovoltaic cells integrated into a building (courtesy of the DOE National Renewable Energy Laboratory).

it to drive a low-temperature gas turbine; focusing the power with an array of mirrors to provide high temperatures for a variety of purposes, including electricity generation; and producing electricity directly using photovoltaic cells. The integration of photovoltaic power into a building is shown in Figure 2.5. This is an attractive use of distributed electricity generation, avoiding transmission losses.

The World Energy Council estimated the total use of solar energy in 1996 to be 3 Mtoe of heat and 0.1 Mtoe of electricity from 330 MW(e) (megawatts of electricity) of thermoelectric power and 138 MW(e) of photovoltaic power.[25] Interestingly, although currently expensive when compared to electricity on the grid, solar-produced electricity can be a cost-effective solution at small levels in areas far from a grid. As prices drop, its use will increase substantially.

It is difficult to set a level for the potential of solar power exploitation since, in principle, it could provide for all energy needs if developments lower the costs sufficiently. The value shown in Table 2.4 (4,000 Mtoe/a in electrical energy) is derived from a case in which 0.1 percent of the world's land area, excluding Antarctica, is covered with photovoltaic collectors generating an average of 40 W(e)/m^2—for example, at 200 W/m^2 and 20 percent photoelectric conversion efficiency. (As a figure for comparison, about 0.1 percent of the land area in the

United States is covered with buildings.) In the final section of this chapter, which compares potential energy demand with potential supply, it is assumed that solar power will play an important role in countries with high insolation, particularly to meet distributed energy needs and possibly to produce hydrogen.[26]

OCEAN THERMAL CONVERSION

Ocean thermal conversion uses the approximately 20°C temperature difference between surface water and deeper water (about 1,000 meters) to generate electricity. The cold water also may be used for cooling in hotter regions, and the system may be used to produce freshwater for horticulture and marine culture.[27] To date, very little power has been produced by this method, although there are various studies for its exploitation, notably for island nations.

TIDAL POWER

About 200 terawatt-hours per year (TWh/a), or 17 Mtoe/a, is estimated to be economically recoverable from tidal power; less than 0.6 TWh/a is presently being used.[28] (A terawatt is a million million watts.) Opportunities for exploiting tidal power are limited to a few areas of the world where tides are large. The largest tidal power plant, the Rance Estuary plant in France, has a capacity of 240 MW and an annual output exceeding 0.5 TWh. One novel approach to exploiting tidal power that may increase deployment of this resource uses a dam that closes on itself—a *coffer dam*—to avoid putting a barrier across an estuary.[29]

WAVE POWER

The wave power resource in deep water (deeper than 100 meters) is estimated to be about 1012 to 1013 W, or 750 to 7,500 Mtoe/a.[30] In the North Atlantic, wave power could produce about 50 kW/m.[31] As waves move into shallower water, they generally lose energy, but in some regions, wave energy is focused, leading to "hot spots" of increased power per unit length. Today, relatively little of this resource is being exploited—about 1 MW(e)—but a number of systems have been tested.

WIND POWER

Wind power potential is highest in Australia and New Zealand, Eastern Europe and the Russian Federation, Latin America, Mongolia, and the United States. The energy potential of wind power is much larger than the world's present electricity use—as much as 500,000 TWh/a—but restrictions on land use are expected to limit its deployment to about 10 percent of this potential—that is, 53,000 TWh/a, or 4,500 Mtoe/a (e).[32] At the end of 1999, the installed capacity was

Figure 2.6. Photograph of a wind farm (courtesy of the DOE National Renewable Energy Laboratory).

more than 10 GW(e), producing some 21 TWh/a of electricity. Modern wind machines range in size from a few kilowatts to 2.5 MW(e). An example of a wind farm is shown in Figure 2.6.

Nuclear Energy

There are two forms of nuclear energy. Fission energy is released when the isotopes of heavy elements, such as uranium and plutonium, are split by the impact of neutrons. Fusion energy is released when light nuclei, such as deuterium, tritium, and helium-3, are fused following collisions. In both cases the binding energies of the nuclei in the end products are less than in the initial ones. The difference in binding energies is energy released.

FISSION

At the end of 1996, there were 439 nuclear power plants in operation, with a capacity of 350 GW(e), generating 2,300 TWh of electricity. A cutaway diagram of a fission reactor is shown in Figure 2.7. In terms of fossil energy replacement value for electricity production, assuming a 35 percent efficiency for fossil

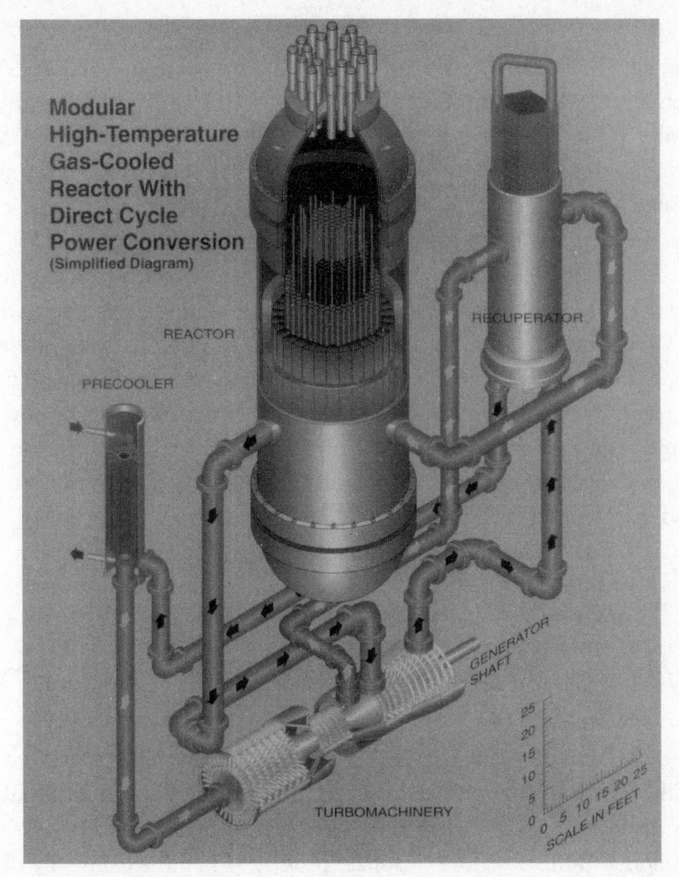

Figure 2.7. Cutaway drawing of an advanced fission reactor (courtesy of J. Simpson, Oak Ridge National Laboratory).

plants, the effective energy was 560 Mtoe, or about 6 percent of world energy use.

At present, most nuclear power plants are operated on a once-through cycle, with disposal of the unused fissionable materials remaining or produced in the fuel. Some recycling of fissionable materials is done in Europe and is planned for Japan and Russia. Even with recycling, the amount of uranium estimated to be available at a low enough cost to make fission power economical (given the present low prices for fossil fuel) is somewhat limited. Annual production of uranium in 1996 was 36,000 tonnes, and demand is expected to double by 2015.[33]

Estimates of recoverable uranium available at a cost of up to $130 per kilogram of uranium include reasonably assured resources of 2.4 million tonnes of uranium (MtU), with another 10.6 MtU expected to be recoverable.[34] Grubler,

Jefferson, and Nakicenovic estimate additional occurrences to total 24.6 MtU.[35] The energy in 1 tonne of uranium (tU) in a once-through cycle is equivalent to 10,000 toe, so 70,000 tU per year amounts to 700 Mtoe/a. If all of the resources are realizable, they would last for about 450 years at this annual rate and could be increased with recycling by about two times and a further one and one-half times by reducing the uranium-235 in tailings to 0.1 percent.

In a breeder reactor, the uranium-238 is converted to fissionable material, and the energy content is 500,000 toe per tonne of uranium. With breeder reactors, this low-cost uranium could supply about 3,000 Mtoe(e)/a of electricity from 15,000 tU/a (40 percent efficiency) for around one thousand years. In fact, with a breeder reactor, uranium cost is less of an issue, and far greater resources of uranium would be economical, possibly including uranium from seawater, so one thousand years would not be the limit of the supply. Large amounts of thorium also exist for use as a breeder fuel.

However, with a greater use of nuclear energy, the issues of public acceptance, the handling of wastes, and weapons-grade-material proliferation remain. It will be important to resolve these issues before fission energy can be used at a massive level.

FUSION

Fusion power is a potentially useful complement to fission power and has advantages with regard to safety, waste disposal, and proliferation issues that would make its deployment advantageous in some situations.[36]

Although a number of fuel cycles are possible, the most likely one uses deuterium and tritium produced by neutron irradiation of lithium. Lithium occurs in substantial quantities and might be extracted from seawater.

A pure deuterium cycle, while more difficult to realize, might also be possible. Deuterium, which occurs naturally as one part in 6,500 of all hydrogen, is effectively limitless. It is estimated that water contains more than 10 million million tonnes of deuterium. The energy equivalent, if all the deuterium and the by-products of its fusion were fused, approaches a million million gigatonnes of oil equivalent energy. Present world energy use is only about 10 gigatonnes of oil equivalent per year (gtoe/a) and may rise to 20 to 30 gtoe/a. Therefore, deuterium-based fusion energy has the potential to supply the world for over a billion years! Put another way, the potential fusion energy in 1 gallon of water is three hundred times the energy in a gallon of gasoline.

Other fusion interactions involving helium-3, lithium, and boron offer the possibility of much lower neutron production but are more speculative—

helium-3 because of the question of its availability on earth and lithium and boron because of the higher temperatures required for releasing net energy.

Although the development of economic electricity from fusion energy is a very challenging task, substantial progress has been made in recent years in both magnetic and inertial fusion research in showing that energetic fusion products behave as expected and in demonstrating some of the key technologies. These successes have led to support of the design and research-and-development studies for the International Thermonuclear Experimental Reactor (ITER) in magnetic fusion and the U.S. National Ignition Facility (NIF) in inertial fusion (under construction).[37] The construction of ITER is awaiting a decision, during the next two or three years, of the international partners—the European Community, Japan, and Russia. Assuming the operation of NIF and ITER or similar facilities, it would be possible to develop fusion energy so that commercial power plants might be operating about the middle of the present century. The most likely initiators of the fusion era are countries that are using substantial fission power, plan to use more, have inadequate indigenous energy resources, and have the technical capability to develop fusion power plants (for example, Japan).

Summary

To understand what might be involved in meeting the potential needs of each area of the world sustainably, it is informative to look at both global and local projections for both sustainable and nonsustainable energy resources. These projections are shown in Table 2.4. An interesting feature of the energy resource data as presented in the table is that there is a large enough number of *types* of resources so that the bottom line of total annual energy resources would not be strongly affected if any one of the individual resources were half or twice as large as the estimates.

The Efficient Use of Energy

Today, much of the energy used is wasted. For example, for a passenger car, typically only 10 percent of the energy of the oil in the ground is used to maintain the car's speed in urban driving (see Figure 2.8). The rest of the energy is dissipated in collecting the oil, transporting and transforming it, using it inefficiently in an engine, and overcoming air and road friction. Because a passenger weighs much less than a car, possibly only 2 percent of the energy is used to move the passenger. If the overall efficiency of energy use were doubled—that is, if 20 percent of oil energy moved a car—then people could use half the energy used today to provide the same services!

Figure 2.8. Ten percent of petroleum from the ground moves the car.

The potential for energy efficiency improvements may be illustrated by the example of opportunities for the United States. A discussion of the U.S. potential for energy efficiency is provided in reports by the U.S. Department of Energy and by Sheffield.[38] There are two main categories of improvements to energy efficiency: electricity production and energy end use.

Efficiency in Electricity Production

Total energy use in the U.S. electricity sector in 1993 was about 750 Mtoe, with 57 percent of the electricity generated from coal, 9 percent from gas, 21 percent from nuclear energy, and 9 percent from hydropower.[39] Overall coal plant efficiency, including the energy costs of pollution controls, had not changed much since the 1950s and was around 30 percent. In many parts of the world, generation efficiency is even lower than in the United States—for example, efficiency for coal-fired plants in China is about 15 percent. Large opportunities for improvement exist.

Modern coal plants with reduced pollution operate at 45 percent efficiency, and systems with greater than 50 percent efficiency are being developed. Some modern aero-derived gas turbine generators have efficiencies of 60 percent, and overall plant efficiency might be improved to 70 percent by the addition of a fuel cell. These improved systems will gradually take over as older plants are replaced. If 66 percent of power produced by coal and gas in 1993 had been generated by the coal and gas systems of the future—with efficiencies of 55 percent and 65 percent, respectively—energy consumption in 1993 would have been reduced from 750 Mtoe to about 500 Mtoe.

Efficiency in Energy End Use

Consider what energy savings might have accrued in the United States in 1993 if the energy for electrical power generation had been reduced as described above and if all of the efficiency improvements in energy end use discussed in notes 38 and 39 had been in place. Table 2.5 provides examples of the potential efficiency improvement in energy end uses by contrasting actual 1993 energy consumption in transportation, buildings, and industry with what that consumption might have been with these energy efficiency improvements. These savings can

Table 2.5. Illustrative reductions in energy end use with implementation of efficiency improvements, 1993

	Energy use (Mtoe)		
Category	Actual 1993	Potential with efficiency improvements	Examples of efficiency improvements
TRANSPORTATION			
Automobiles	240	80	Hybrid-electric, fuel cells, tires Materials, lean burn, direct injection stratified charge Improved diesels, aerodynamics
Light trucks	110	55	
Heavy trucks	90	70	
Air transport	50	25	Weight, aerodynamics, engines
Other transp.	90	< 90	
		-50[a]	Intelligent transportation systems
BUILDINGS			
Space heating	220	85	Solar, heat pumps, windows
Space cooling	85	35	Thermal activation, triple-effect chillers
Lighting	120	40	Advanced fluorescents, sulfur lamps
Water heating	90	75	Replacement of old stock
Food storage	60	20	Improved refrigerators
Other[b]	190	120	Distributed generation (CH&P)
		-25[a]	Sensors, controls, factory building

(*continues*)

Table 2.5. Continued

Category	Energy use (Mtoe) Actual 1993	Energy use (Mtoe) Potential with efficiency improvements	Examples of efficiency improvements
INDUSTRY			
Chemical	130	95	Bioprocesses, biotech, catalysts, materials, separations
Refining	150	85	Inert anodes
Primary metals	65	40	Electric arc furnaces, net-shape manufacturing[c]
Pulp & paper	65	50	Impulse drying, materials
Agriculture	40	30	Improved irrigation and fertilization
Other	330	280	Sensors and controls
		-65[a]	Waste heat pumps, recycling, etc.
Total	2,125	1,135	

[a] Minus numbers represent potential additional energy savings for each area, through the use of intelligent transportation systems, sensors and controls, and waste heat pumps and recycling.
[b] Distributed electricity generation, particularly using combined heat and power (CH&P), could offer substantial savings.
[c] In net-shape manufacturing, energy use is reduced because less machining is required to produce the components.
Sources: Sheffield 1997, DOE-EERE 1997, DOE-EIA 1996.

be substantial. For example, in the United States, the Partnership for a New Generation of Vehicles has the goal of developing a medium-sized automobile with three times better fuel consumption (80 miles per gallon rather than 27) by 2008, with the same safety, comfort, and cost as today's models.[40] Similar efforts are under way in Europe and Japan.

The rough estimates of savings shown in Table 2.5, together with electricity generation improvements, illustrate how, with present and projected developments of the next decade or so, it should be possible to use half as much energy to achieve the same benefits. In most of the developing countries, energy is used far less efficiently today than in the United States and other developed countries. Thus, even greater efficiency improvements are possible—for example, in the case of steel production in China.[41] In fact,

Table 2.6. World population projection by region (millions), 2010-2150

Region	2010	2025	2050	2075	2100	2125	2150
N. Amer	334	360	375	380	385	385	385
L. Amer	585	680	820	905	940	940	940
Eur. OECD	464	465	460	460	455	455	455
FSU + CEE	443	460	490	500	515	515	515
Pac. OECD	154	150	140	140	135	135	135
China	1,348	1,495	1,640	1,695	1,695	1,695	1,695
East Asia	753	860	1,005	1,095	1,135	1,140	1,140
South Asia	1,557	1,915	2,385	2,640	2,785	2,845	2,850
Africa	1,055	1,360	1,865	2,250	2,725	2,975	3,115
Mid. East	237	335	445	535	560	560	560
Total	6,930	8,080	9,625	10,600	11,330	11,645	11,790

Source: Sheffield 1998.

many experts believe that the world could use three or even four times less energy for the same benefits if all avenues for improving efficiency were exploited.

Future Energy Demands and Resources

It is important to understand, region by region, how the energy situation might evolve over the next century. As shown in Table 2.2. and Figure 2.4, there are a variety of projections. Consistent with the energy topic of this book, we present an example in which the annual per capita energy use is related to the population growth rate assuming that the historic trends shown in Figure 2.3 continue (see Appendix 3).[42] Improvements in the efficiency of energy use (75 percent improvement by 2100) and modest cultural changes were invoked in this projection of coupled energy use and population growth. No reference was made to actual energy resources. A comparison of potential supplies to meet the demand is given below.

For the period from 1995 to 2010, the population projections of the World Bank and the commercial energy per capita given by the International Energy Agency were used.[43] Biomass energy estimates given by studies for the United Nations were included on the assumption that biomass energy would be used

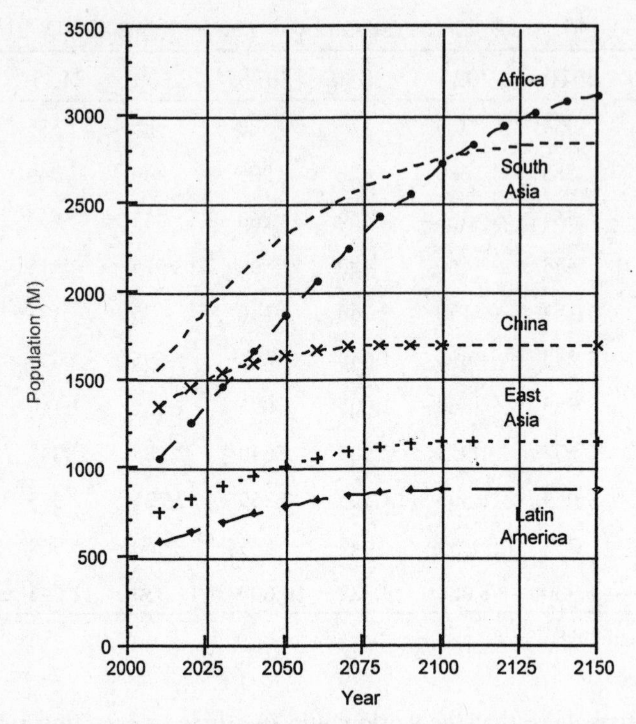

Fig. 2.9 Hypothetical case: projected population in the developing areas, 2010–2150.

more effectively in the future and, therefore, should be treated as a commercial energy.[44]

This approach led to population changes following closely the projections of the World Bank for the period up to 2150, and a relatively low demand for energy, compared to other projections. The hypothetical populations for the various areas of the world are shown in Table 2.6 and for the developing areas in Figure 2.9.

Table 2.7. World energy projection by region (Mtoe/a), 2010-2150 (Hypothetical case: 75% efficiency gain by 2100)

Region	2010	2025	2050	2075	2100	2125	2150
N. Amer	2,679	2,895	3,000	3,055	3,080	3,080	3,080
L. Amer	782	945	1,190	1,365	1,415	1,355	1,300
Eur. OECD	1,754	1,750	1,740	1,730	1,720	1,720	1,720
FSU + CEE	1,621	1,690	1,790	1,840	1,880	1,880	1,880
Pac. OECD	793	770	730	715	700	700	700
China	1,521	1,960	2,230	2,240	2,030	1,940	1,860
East Asia	931	1,100	1,345	1,500	1,510	1,490	1,425

Region	2010	2025	2050	2075	2100	2125	2150
South Asia	825	1,410	2,375	2,915	2,885	2,995	2,900
Africa	615	850	1,315	1,830	2,320	2,795	3,145
Mid. East	519	770	975	1,070	1,010	970	930
Total	12,040	14,140	16,690	18,260	18,550	18,925	18,940

* East Asia includes Afghanistan, Bhutan, Brunei, Indonesia, North and South Korea, Laos, Malaysia, Myanmar, Papua/New Guinea, Philippines, Singapore, Taiwan, Thailand, Vietnam, and various Pacific islands (IEA definitions).
* South Asia includes Bangladesh, India, Nepal, Pakistan, and Sri Lanka (IEA definition).
Source: Sheffield 1998.

Annual Energy Use by Region

Projected annual energy use is shown in Table 2.7 and for the developing parts of the world in Figure 2.10. World energy use was projected to rise from about 9,000 Mtoe/a today to 19,000 Mtoe/a by the time the world's population has risen from 6 billion to around 11 billion people in the twenty-second century. As can be seen from a comparison with the examples in Table 2.2, this hypothetical case is on the optimistic side: if there were no improvements in energy efficiency and no "cultural" changes, and there were a continuing increase in per capita energy use in the developing world, the energy demand would be over

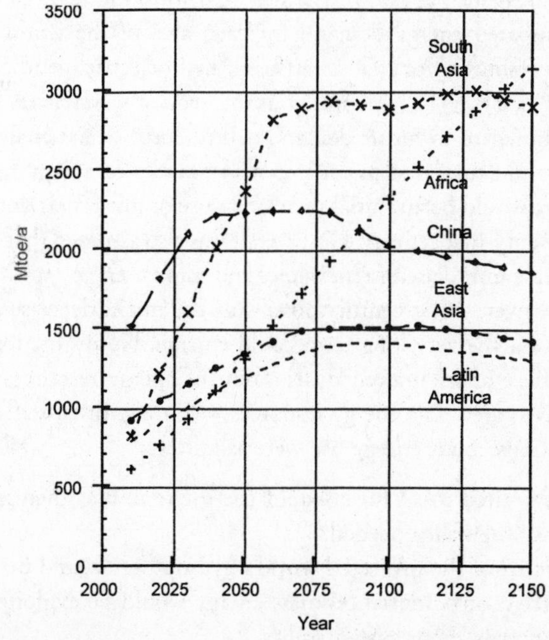

Fig. 2.10 Hypothetical case: project energy use in the developing regions, 2010–2150.

40,000 Mtoe/a. The problems of having a distribution of energy that does not match the forecasted needs for some regions is discussed below.

Meeting Future Energy Demand

All projections for world energy use posit a steadily increasing demand in each area of the world. This gives rise to two key questions:

- How might this increasing energy demand be met?
- What energy resources are available in each part of the world?

These are issues even for the optimistic hypothetical case discussed above and certainly for the other projections shown in Table 2.2, even though the world's resources appear to be large enough (see Table 2.4).

One problem is that in some areas, indigenous energy resources do not match projected needs. Of course, this is already true today for some countries—for example, Japan. But it has never been an issue at the scale projected for the future. Japan has a population of only about 125 million people, each using a few tonnes of oil equivalent per year. On the other hand, the population of South Asia (the Indian subcontinent) is projected to rise to 2.5 to 3 billion people, needing more than 1 toe per person, even with efficiency improvements. As we shall see below, the demands of areas such as South Asia will be hard to meet.

Sheffield compared energy demand for each area of the world against the potential energy resources for that area (fossil, hydroelectric, geothermal, wind, biomass, nuclear, solar).[45] The remainder of this section summarizes the findings of that analysis in regard to future demand and resources. That analysis assumed that countries would turn first to indigenous resources of energy to meet their needs; thus, there would be an initial increase in use of fossil fuels during the first decades of the twenty-first century. However, it was also assumed that there would be a steady increase in the use of renewables and nuclear energy so that, from the latter part of the twenty-first century, fossil fuel use might decrease and a stable energy situation for the very long term could emerge. No distinction was made regarding the effect of an uneven distribution of energy resources among the countries of a given area. The energy resources were those discussed above.

The assumptions about energy use were as follows:

- Regions with limited fossil fuels would use them all and then import what additional fossil fuels they needed.
- About 90 percent of the projected world's hydropower would be exploited.
- About 50 percent of potential biomass energy would be exploited, a limitation set by concerns about sustainability.[46]

- About 40 percent of the wind power potential would be used, consistent with constraints on land use and need.[47]
- The balance of energy for each region would be provided mainly by some combination of nuclear and solar energy, with modest amounts of geothermal, tide, and wave power energy included in favorable situations.

Electrical Energy Use

The past trend toward an increasing use of electrical energy is expected to continue.[48] The increase may even accelerate if electricity-producing energy sources—such as hydro, nuclear, solar electric, wind, wave, and tide—are used to replace fossil fuels. A question arises as to how to treat electrical energy in estimates of energy demand and energy resources: as a primary energy source or as a replacement value for fossil energy? For hydropower, the International Energy Agency (IEA) defines energy use only as the electrical energy produced, but in discussing electricity from nuclear and fossil energy, it uses the energy contained in the fuel.

Today, most electricity is produced from fossil fuels, at 20 to 30 percent average efficiency. In the discussion below, the comparison of demand and resources for the future assumes that alternatives to fossil energy are replacing fossil energy and that their equivalent thermal energy will be used. However, the efficiency of fossil electricity production is increasing steadily and the waste energy is decreasing; therefore, allowing for future improvement, an effective efficiency of 50 percent will be assumed after 2010. The improvements in the efficiency of electricity end use and cogeneration heat use are accounted for in a blanket improvement in energy efficiency assumed for the hypothetical case.

There is an increasing move to using distributed electricity energy, from biomass, small hydropower, solar and wind power, and the availability of more natural gas, rather than centralized sources. This move has been facilitated by improvements in gasifiers, solar cells, and microturbines.

Indigenous Energy Resources in Developing Regions

Table 2.8 shows the results of an assessment of the ability of each region of the developing world to meet the projected demands of the hypothetical case for the year 2100 through the use of its indigenous energy resources.

To arrive at the figures shown in this table, first the expected *sustainable* amounts of the generally large, renewable resources of hydropower, wind, and biomass energy were subtracted from the total demand. Any shortfall in energy needs must then be met by some combination of fossil, nuclear, and solar energy.

Table 2.8. Projected annual energy use (Mtoe/a) for developing regions in 2100 compared with estimated indigenous resources

	Africa	China	East and South Asia	Latin America
PROJECTED DEMAND, 2100	2,320	2,030	4,395	1,415
Indigenous sustainable resources				
Hydro + wind power	980	320	670	330
Biomass	1,040	490	560	1,540
Shortfall	300	1,220	3,165	0
Indigenous fossil resources				
Fossil for 100 years[a]	1,120	2,260	1,330	1,070
(Present fossil production[b])	(550)	(950)	(650)	(550)
Shortfall	0	0	1,835	0
Nuclear and solar resources				
Fission (LWR) for 100 years[c]	280	180	30	90
Solar: 0.1% of land area[d]	800	290	270	600
Shortfall	0	0	1,535	0

[a] Estimated recoverable resources of oil, gas, coal, shale oil and bitumen (World Energy Council, *Survey of Energy Resources,* 1995, 1998). Larger amounts of more speculative resources may exist.

[b] World Energy Council, *Survey of Energy Resources,* 1995, 1998.

[c] Once-through fission cycle in light water reactor (LWR). With breeder reactors, the energy available would increase by fifty times. Note also that South Asia has large thorium resources.

[d] Collection over 0.1% of land area at 40 W_e/m^2 average solar power.

Sources: Sheffield 1998, WEC, 1995 and 1998 *Survey of Energy Resources,* London: World Energy Council. Johansson, T. B., H. Kelly, A. K. N. Reddy, and R. H. Williams, eds.1993. *Renewable Energy: Sources for Fuels and Electricity.* Washington, D.C.: Island Press. Larson, E. D., and R. H. Williams. 1995. "Biomass Plantation Energy Systems and Sustainable Development." In J. Goldemberg and T. B. Johanson, eds., *Energy as an Instrument for Socio-Economic Development.* New York: U.N. Development Programme.

Possible indigenous fossil energy use for the next one hundred years to meet additional needs can then be compared to today's annual production (to put future demand in perspective). Any remaining shortfall must then be met by nuclear and solar energy or by imported energy. For convenience, energy resources with lower potential (such as geothermal, tide, and wave power) were included under the solar category. While these resources certainly have importance for some countries, they generally are not a large part of a region's resources.

No attempt was made to assess the distribution of energy resources within a region. Clearly, some countries may be very deficient in energy resources when compared with others—for example, Somalia compared with Nigeria, Peru compared with Brazil, or Bangladesh compared with Pakistan.

The main shortfall problems seem to lie in East and South Asia, even if fossil resources were double the World Energy Council's current base estimate. The shortfall projected for East and South Asia in this hypothetical case is comparable to the total energy use in the United States today. Africa, while it has good energy resources, may suffer from the resources' uneven distribution and less of an infrastructure to capitalize on them. The populations of these areas are already huge, their energy use per capita is relatively low, and their growth rates are high. Unless there are much greater growth reductions through "cultural" factors than are assumed here, these regions face problems if the energy per capita model is more or less right. And whether the assumptions in these calculations are precisely right or not, they give an indication of the kind of changes that will be needed if the world's population is to be stabilized at some sustainable level. Ideally, as shown in the example, those regions with shortfalls should rapidly boost their energy usage and stabilize their populations.

Importation of Energy

In the 1990s, many countries imported a high percentage of their commercial energy. Only in the developed countries, however, was the annual imported energy per capita high. For example, in Western Europe, energy imports were in the range of 1.4 to 4.8 toe/cap·a in 1991.[49] By contrast, in the developing countries, while imports represented a high percentage of commercial energy, the levels of imports per capita was low (0.012 toe/cap·a for China, 0.015 to 0.10 toe/cap·a for Sub-Saharan Africa and South Asia, 0.018 to 0.36 toe/cap·a for East Asia, and 0.031 to 0.53 toe/cap·a for Latin America): thus, to meet projected future shortfalls, East and South Asia and the energy-poor developing countries not discussed here will need substantial economic gains to be able to afford a large importation of energy.

The distribution of energy use by source derived for this hypothetical case is shown in Figure 2.11. The category "Other" at the top of the figure refers to additional energy sources needed in the developing countries beyond the nominal levels of indigenous resources—for example, a greater use of solar and nuclear power and a higher level of imported fossil fuels. The distribution shown in this graph is similar to numerous other projections. For example, in terms of fossil fuel use, the case shown is in the range between the IPCC's cases IS92c and IS92d.[50] The ultimate reduction in greenhouse gas emissions, particularly if there is also some carbon sequestration, is also consistent with the IPCC's goal of reducing greenhouse gas emissions.

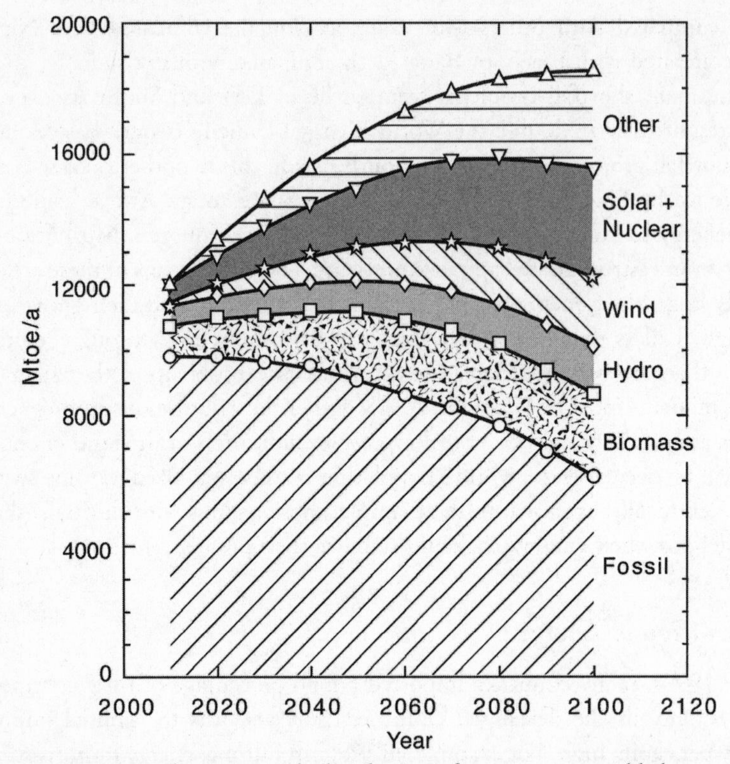

Fig. 2.11 Hypothetical case: example distribution of energy sources added to meet the reference world energy demand for 2010–2100. This case emphasizes efficiency improvements and growth in nuclear and renewable energies. Carbon sequestration may be needed to meet reference IPCC goals while using this amount of fossil fuel.

During the transition from the present fossil fuel–dominated situation to one involving more renewable energies, cheap fossil fuel–derived imports are probably essential to facilitate a smooth transition in the developing regions to a stable population with an improved standard of living. Issues related to more massive fossil fuel use by developing countries include the availability of cheap, mobile fuel and, as noted above, the ability of the developing countries to purchase it. Today, most of the imported fuel is oil, and as conventional oil reserves are depleted, the question arises as to what could be substituted as a readily transportable fuel: oil from unconventional resources? gas? carbon-based gas or liquid fuels? hydrogen from coal and biomass and electricity? A renewables-intensive global energy scenario is described by Johansson et al. for the next sixty years, using large amounts of methanol and hydrogen from biomass.[51] Production of hydrogen using solar energy is also discussed in this reference.

The renewables-intensive global scenario uses a diversity of energy sources, the relative abundance of which varies from region to region in a similar manner to the example discussed earlier. By 2050, 60 percent of the electricity would be provided from renewable energies in a combination of hydropower and intermittent sources (wind, solar thermal-electric, solar photovoltaic), biomass, and geothermal power. At the same time, 40 percent of fuels would be provided by methanol, ethanol, hydrogen, and methane (biogas) derived from biomass. Hydrogen fuel would also be provided by electrolysis of water, using intermittent renewable electricity. The remaining energy would be provided by fossil fuels and nuclear fission. Emphasis would be placed on the efficient use of energy.

What Might the Future Look Like? Consequences of Limited Per Capita Energy Growth

An interesting exercise in showing the potential problems resulting from low annual per capita energy use in a particular region is to take the simple formula discussed in Appendix 3, which relates population growth rate to the annual useful energy used per capita, and compare two cases for total energy use in a developing region over a fixed period of time.

Case 1: Let the energy use per capita be raised immediately to the level at which the population growth rate is zero.

Case 2: Let there be no change in energy per capita use in the region.

How long does it take for the total energy use to be the same for the two cases, and what are the differences in population, assuming a start with 1 billion people?

After about 160 years, Case 1—the case with the zero population growth rate and the higher per capita energy use—will still have 1 billion people. However, in Case 2, where there is no change in per capita energy use, the population will have increased to about 16 billion people! This result is not sensitively dependent on the actual level of per capita energy use provided it is less than the critical amount at which the population growth rate becomes zero.

Clearly, such population growths, at the present energy use of 0.6 to 0.8 toe/cap·a in the developing world, would be intolerable if they were to continue; they illustrate the point that the world does as well or as badly as the region with the highest growth rate. Cultural advances may sufficiently ameliorate the problem by lowering fertility rates in those areas that do not get as much energy as is needed to otherwise stabilize population. These simple calculations also make the point that the world as a whole would benefit from an increase now in the standard of living (annual energy per capita), because in the end—about 140 to 180 years hence—such an increase might make little difference to the total demand for energy over that period but it would make an enormous difference to its sustainability.

Although it may be possible for a developing region to decrease its population growth rate while not increasing its annual per capita energy use, the trends shown in Figure 2.3 suggest that, on average, this is an improbable result; although some rapid reductions in fertility rate have occurred, they may not always be sustained.[52] It seems unreasonable to expect the developing regions to make such changes independent of improvements in standards of living, which imply some increased energy use. Nevertheless, one factor that should allow a reduction of energy requirements is the introduction of efficiency improvements, which can improve standards of living at any energy level. Even allowing for efficiency improvements, an increase in annual energy use per capita for the poorer countries would be a desirable occurrence from a humanitarian viewpoint. A cursory look at the energy resources discussed above supports the conclusion that not only substantial efficiency improvements but also the use of all forms of energy will be needed to meet the world's energy needs at a level sufficient to stabilize world population over the next one hundred years.[53]

If the apparent connection between the fertility rate and per capita energy use holds in the future, then it indicates a problem for the developing areas: *stabilizing their populations will require more energy per capita now.* Failure to raise the per capita energy use can lead to sustained high rates of growth of population and, therefore, an even higher energy use later. It is not clear whether "cultural" changes can always work rapidly enough to lower the rate and to stabilize population without the accompanying improvements in standard of living bound up with increased energy use, though Horiuchi cites the examples of a rapid fertility decrease in China (6.45 children per woman in 1968 dropping to 2.24 children in 1980) and a more moderate rate in India, and Cohen cites examples of rapid fertility reductions in Colombia and Thailand.[54]

Conclusions

The major challenges in obtaining energy to raise the standard of living and to stabilize the population appear to be for Africa, East Asia, and South Asia. The population in these regions is already large and is growing rapidly. If the dependence on energy use per capita is a fair measure of the standard-of-living aspects of population growth, the countries in these areas need to raise their energy use rapidly enough to stabilize their populations before energy demand reaches untenable levels. Of course, if the "cultural" aspects of population growth reduction can be realized more rapidly, these problems can be ameliorated.

It appears that the availability of easily movable, cheap fuel is essential for the developing areas to increase their energy use, improve their standards of living, and stabilize their populations at a sustainable level. In the near term, fossil fuels can fulfill this role, within constraints set by environmental considerations. In the longer term, a sustainable solution requires the development and deployment of all energy sources to complement and ultimately replace the fossil fuels.

Two extreme possibilities for the future are (1) that the world continues to use its fossil reserves rapidly and (2) that concerns about fossil usage slow down the rate of fossil use dramatically. The former scenario could pose a problem for developing countries because the rapid use of fossil fuels may limit these countries' ability to obtain for a sufficient length of time lower cost, easily moved fossil fuels while they make the transition to renewable and nuclear energies. The latter scenario is the more desirable one because it should lengthen the time of availability of cheaper fossil fuel. On the other hand, it could also limit the use of such fuel by developing countries.

It seems probable that there are far greater resources of fossil fuels than are presently classified as readily recoverable. One can hope that more oil and gas will be found and that technological advances will make possible the recovery of fossil fuels beyond the amounts currently proven and estimated. However, unless such finds represent a huge increase over that used and projected, the existence of fossil energy would not alter the simple point that at some time, whether it be 2100 or 2150 or 2500, the world must face up to providing most of its energy through sources other than fossil fuels. A consequence of this situation is that all energy sources as well as a vastly improved efficiency of energy use will be needed to provide a sustainable "decent" future for the people of the world.

Key Ideas in Chapter 2

- World population is increasing. Per capita energy use is an important measure of standard of living, and raising the level in developing countries seems to be essential. Combined, these factors lead to a substantial need for more energy.

- We are entering a multienergy era. A broad portfolio of energy resources and technologies will be needed by the middle or the end of the twenty-first century to supplement and ultimately replace nonrenewable fossil fuels.

- The world has huge energy resources, but they are unevenly distributed.
- Meeting future world energy demand will require significant improvements in the efficiency of energy use, substantial reductions of wasted energy, and stabilization of population and energy use per capita.
- Areas most in need of more energy to raise the standard of living and help stabilize the population appear to be Africa, East Asia, and South Asia.

Acknowledgment

I appreciate very much the contributions of Carolyn Moser of the Oak Ridge National Laboratory to editing and improving the quality of this chapter.

Notes

1. While the focus here is only on opportunities and issues related to energy supply and demand, there are similar concerns about the supply of clean water and adequate energy and water to provide increasing amounts of food. It is important to realize that if we are to have a sustainable future, there cannot be both a permanent growth in population and a continuing growth in the use of nonrenewable energies and other resources. Therefore, recycling of materials and the efficient use of resources is crucial to meeting the world's future needs. H. Daly, "Sustainable Growth—An Impossibility Theorem," *Development* 3:4 (1990): 45; A. A. Bartlett, "Reflections on Sustainability, Population Growth, and the Environment," *Population and Environment: A Journal of Interdisciplinary Studies* 16:1 (1994): 5–35.
2. J. Goldemberg and T. B. Johansson, eds., *Energy as an Instrument for Socio-Economic Development* (New York: United Nations Development Programme, 1995).
3. J. Goldemberg et al., "Basic Needs and Much More with One Kilowatt per Capita," *AMBIO: A Journal of the Human Environment* 14:4 (1994): 190–200.
4. E. Bos et al., *World Population Projections: 1994-95 Edition* (Baltimore: Johns Hopkins University Press, for the World Bank, 1994). United Nations, *World Population Prospects: The 1992 Revisions* (New York, 1993).
5. J. E. Cohen, *How Many People Can the Earth Support?* (New York: Norton, 1995).
6. J. P. Holdren, "Energy in Transition," *Scientific American* 263 (1990): 157–63; S. W. Gouse et al., "Potential World Development through 2100: The Impacts on Energy Demand, Resources, and the Environment," *World Energy Council Journal* (December 1992): 18–32; J. Sheffield, "Population Growth and the Role of Annual Energy Use Per Capita," *Technological Forecasting and Social Change* 59:1 (1998): 55–87.
7. F. Duchin, "Global Scenarios about Lifestyle and Technology," Paper presented at the Sustainable Future of the Global System, United Nations University, Tokyo, October 1995.
8. Sheffield, "Population Growth."
9. United Nations, *Statistical Yearbook,* 10th issue (New York: U.N. Publishing Divi-

sion, 1965); *Statistical Yearbook,* 20th issue (1975); *Statistical Yearbook,* 22nd issue (1977); *Statistical Yearbook,* 32nd issue (1987); *Statistical Yearbook,* 39th issue (1994).

10. A. Sen, "The Economics of Life and Death," *Scientific American* 268 (May 1993): 40–47.
11. B. Robey, S. O. Rutstein, and L. Morris, "The Fertility Decline in Developing Countries," *Scientific American* 269 (December 1993): 60–67.
12. United Nations, *Statistical Yearbook* (1965, 1975, 1977, 1987, 1994).
13. Goldemberg et al., "Basic Needs."
14. World Energy Council, *1995 Survey of Energy Resources* (Oxford, Eng.: Holywell Press, 1995); A. Grubler, M. Jefferson, and N. Nakicenovic, "Global Energy Perspectives: A Summary of the Joint Study by the International Institute for Applied Systems Analysis and the World Energy Council," *Technological Forecasting and Social Change* 51:3 (1996): 237–64.
15. M. D. Max, R. E. Pellenbarg, and B. G. Hurdle, *Methane Hydrate, a Special Clathrate: Its Attributes and Potential,* NRL/MR/6101-97-7926 (Washington, D.C.: Naval Research Laboratory, February 28, 1997).
16. T. M. L. Wigley, R. Richels, and J. A. Edmonds, "Economic and Environmental Choices in the Stabilization of Atmospheric CO_2 Concentrations," *Nature* 379 (1996): 240.
17. International Panel on Climate Change, *Climate Change, 1995: Impacts, Adaptations, and Mitigation of Climate Change—Scientific and Technical Analyses,* ed. R. T. Watson et al. (Cambridge: Cambridge University Press, 1996).
18. W. C. Turkenburg, "Sustainable Development, Climate Change, and Carbon Dioxide Removal (CDR)," *Energy Conversion Management* 38 (1997): Suppl. S3–S13; H. Herzog, E. Drake, and E. Adams, *CO_2 Capture, Reuse, and Storage Technologies for Mitigating Global Climate Change,* final report for DOE Order No. DE-AF22-96PC01257 (Cambridge: Massachusetts Institute of Technology Energy Laboratory, January 1997); R. Socolow, ed., *Fuels Decarbonization and Carbon Sequestration,* PU/CEES no. 302, report of a workshop by the members of the report committee, Center for Energy and Environmental Studies, Princeton University, Princeton, New Jersey, 1997.
19. M. J. Grubb and N. I. Meyer, "Wind Resources," in T. B. Johansson et al., eds., *Renewable Energy: Sources for Fuels and Electricity* (Washington, D.C.: Island Press, 1992).
20. D. O. Hall et al., "Biomass for Energy: Supply Prospects," in Johansson et al., eds., *Renewable Energy;* E. D. Larson and R. H. Williams, "Biomass Plantation Energy Systems and Sustainable Development," in Goldemberg and Johansson, eds., *Energy as an Instrument for Socio-Economic Development.* (New York: United Nations Development Programme, 1995).
21. D. Pimentel, "Biomass Fuel and Environmental Sustainability in Agriculture," in T. B. Mepham, G. A. Tucker, and J. Wiseman, eds., *Issues in Agricultural Bioethics* (Nottingham, U.K.: Nottingham University Press, 1995); D. Pimentel, "Environmental and Economic Costs of Soil Erosion and Conservation Benefits," *Science* 267 (February 1995): 1117–24.

22. C. G. Palmerini, "Geothermal Energy," in Johansson et al., eds., *Renewable Energy.*
23. World Energy Council, *1998 Survey of Energy Resources* (London: World Energy Council, 1998).
24. World Energy Council, *1995 Survey of Energy Resources.*
25. World Energy Council, *1998 Survey of Energy Resources.*
26. R. Socolow, ed., *Fuels Decarbonization and Carbon Sequestration;* Johansson et al., eds., *Renewable Energy.*
27. World Energy Council, *1998 Survey of Energy Resources.*
28. World Energy Council, *1998 Survey of Energy Resources.*
29. Personal communication from P. Ullman and J. Gault, John Gault SA, Geneva, Switzerland, 1997.
30. World Energy Council, *1995 Survey of Energy Resources;* World Energy Council, *1998 Survey of Energy Resources.*
31. J. E. Cavanaugh, J. H. Clarke, and R. Price, "Ocean Energy Systems," in Johansson et al., eds., *Renewable Energy,* 513.
32. Grubb and Meyer, "Wind Resources"; B. Sorensen, "History of, and Recent Progress in, Wind-Energy Development," in *Annual Review of Energy and the Environment* (Palo Alto, Calif.: Annual Reviews, 1995) 20: 387–424.
33. World Energy Council, *1995 Survey of Energy Resources;* World Energy Council, *1998 Survey of Energy Resources.*
34. World Energy Council, *1998 Survey of Energy Resources.*
35. Grubler, Jefferson, and Nakicenovic, "Global Energy Perspectives."
36. S. Barabaschi (chair), *Fusion Programme Evaluation,* Euratom Report 17521, European Commission DG-XII, Brussels, Belgium, 1996.
37. International Thermonuclear Experimental Reactor, *Technical Basis for the Final Design Report, Cost Review, and Safety Analysis (FDR)* (Vienna: International Energy Agency, 1998); National Ignition Facility, *Conceptual Design Report,* UCRL-PROP-117093 (Livermore, Calif.: Lawrence Livermore National Laboratory, 1994).
38. U.S. Department of Energy, Office of Energy Efficiency and Renewable Energy, *Scenarios of U.S. Carbon Reductions: Potential Impacts of Energy-Efficient and Low-Carbon Technologies by 2010 and Beyond,* ORNL/CON-444, report coordinated by Lawrence Berkeley and Oak Ridge National Laboratories in conjunction with Argonne, National Renewable Energy, and Pacific Northwest National Laboratories (Oak Ridge, Tenn.: Oak Ridge National Laboratory, 1997); J. Sheffield (coordinator), *Energy Technology R&D: What Could Make a Difference?* vol. 2, pt. 1, *End-Use Technology,* ORNL-6921/V2/P1 (Oak Ridge, Tenn.: Oak Ridge National Laboratory, 1997).
39. U.S. Department of Energy, Energy Information Administration, *Monthly Energy Review,* DOE/EIA-035(96/11) (Washington, D.C.: U.S. Department of Energy, Energy Information Administration, November 1996).
40. Partnership for a New Generation of Vehicles, *Program Plan: Declaration of Intent* (Dearborn, Mich.: U.S. Council for Automotive Research, September 1993); Partnership for a New Generation of Vehicles, *Program Plan* (Dearborn, Mich.: U.S. Council for Automotive Research, July 1994).

41. E. Worrell, "Advanced Technology and Energy Efficiency in the Iron and Steel Industry in China," *Energy for Sustainable Development* 2 (1995): 27–40.
42. Sheffield, "Population Growth."
43. Bos et al., *World Population Projections: 1994–95 Edition;* International Energy Agency, *World Energy Outlook, 1995 Edition (*Paris: OECD Publications, 1995).
44. Hall et al., "Biomass for Energy"; Larson and Williams, "Biomass Plantation Energy Systems."
45. Sheffield, "Population Growth."
46. Hall et al., "Biomass for Energy"; Larson and Williams, "Biomass Plantation Energy Systems"; Pimentel, "Biomass Fuel and Environmental Sustainability"; Pimentel, "Environmental and Economic Costs."
47. International Energy Agency, *World Energy Outlook.*
48. International Energy Agency, *World Energy Outlook.*
49. United Nations, *Statistical Yearbook* (1994).
50. International Panel on Climate Change, *Climate Change, 1995.*
51. Johansson et al., eds., *Renewable Energy.*
52. S. Horiuchi, "Stagnation in the Decline of the World Population Growth Rate during the 1980s," *Science* 257 (1992): 761–67.
53. J. Goldemberg, "Energy Needs in Developing Countries and Sustainability," *Science* 269 (1995): 1058–59.
54. Horiuchi, "Stagnation"; Cohen, *How Many People Can the Earth Support?*

Chapter 3

Environmental Impacts of Energy Use

RUSSELL LEE

Although considerable uncertainty exists about future needs for energy, it is clear that meeting the energy requirements for sustainability over the next century is going to be a serious challenge. Furthermore, meeting this challenge of providing an adequate energy base is a solution to only half of the energy problem. The other half concerns the environmental impacts of energy use.

Several so-called "clean energy sources" exist, but this perception is correct only in a relative sense. All energy transformations have environmental consequences. This is an inescapable implication of the laws of thermodynamics (energy conservation and entropy increase) discussed in Chapter 1. For example, suppose work is accomplished by burning a gallon of gasoline. Not only is the energy conserved (the first law of thermodynamics) and dissipated (the second law of thermodynamics) but so are the chemicals in altered form. These dissipated residuals include unburned hydrocarbons, such as carbon monoxide and carbon dioxide; water; sulfur dioxide; and various nitrogen oxides. Most of these residual chemicals have at least some deleterious effects on the environment. We call these residuals pollution, *a term that refers to anything that diminishes the value, utility, or efficiency of our natural life-support system on the earth. Some of these residuals may demonstrate their effects only over substantial periods of time (for example, acid rain, greenhouse gases, and nuclear wastes) and consequently are particularly hard to address.*

Can future world energy needs be met without doing intolerable and irreparable damage to the environment? In this chapter, Russell Lee addresses the environmental impacts of energy use, providing a consistent comparison of the environmental effects of alternative technologies that are, or may become, important in meeting our energy requirements.

—Editors' note

Energy is needed to carry out our daily activities in commerce, transportation, industry and, of course, at home. As discussed in Chapter 2, the world's rapidly growing population presents a daunting challenge. How can we provide the energy resources required to sustain and advance our productivity and well-being?

The challenge is even greater than merely providing adequate supplies of energy. The chemical and physical processes whereby energy resources are converted into more useful forms such as electricity result in pollutant emissions and other discharges and burdens on the environment. These pollutants adversely affect both ecosystems and human health. In some instances, these impacts are long-term and irreversible.

Because the various sources and technologies for energy conversion can lead to different types and magnitudes of environmental impacts, it is important to compare them in a clear and consistent way. Such information is indispensable to the public interest in developing energy policy and in setting regulations, standards, taxes, and fees on energy use. Also, companies sometimes assess the possible environmental impacts of their current or planned operations. They might want to identify cost-effective pollution prevention options, possibly in anticipation of future government regulations. Some companies recognize that environmental impacts might ultimately affect the economic health of regions, which will in turn affect the demand for these companies' products and services. In addition, some consumers are willing to pay more for their electricity if they know that it is generated using "environmentally friendly" technologies.

No energy source or technology is entirely benign—energy conversion and use, in whatever form, affects the environment, though in different ways. Some energy sources and technologies are usually more environmentally friendly than others, but we should not make sweeping generalizations that some are "good" or "clean" and that others are "bad" or "dirty." Even for a given technology, there can be significant variations in the environmental impacts. Environmental impacts should be estimated using sound scientific methods, and it is necessary to know the more severe impacts for different types of energy sources and technologies that are used to generate electricity. Unfortunately, the more environmentally benign energy sources are not used as often as they could be.

Energy Conversion

Energy conversion is needed to produce useful forms of energy. Consider, for instance, electricity, which is widely used for lighting, computer operations, scientific and industrial equipment, household appliances, and many other things.

Electric power is the product of *current* (the movement of electrons) and *voltage* (electrical potential) and is obtained through the conversion of other forms of energy, such as thermal, mechanical, or chemical.

Chemical energy in fossil fuels—principally, coal, natural gas, and oil—is converted into electric power by burning the fuel. The resulting heat is combined with water to produce steam, which drives turbines that are coupled to electric generators which, in turn, convert the mechanical energy from the turbine into electricity. Chemical energy is also stored in biomass such as wood, which can be burned to generate electricity as well.

Electricity can also be generated by water and wind. In hydropower, a water turbine uses the gravitational potential energy stored in a reservoir. The potential energy lies in the difference in elevation between an upstream water reservoir and the point at which the water exits the turbine. The falling water rotates the turbines, which convert the potential energy into mechanical energy. The turbines in turn drive generators, which convert the turbines' mechanical energy into electricity. In the case of wind energy, wind turbines extract the kinetic energy in the wind. The wind passes through the turbines to create mechanical power for an electric generator.

In nuclear reactors, the fission of heavy atomic nuclei releases energy. The products of the fission process are expelled at high speeds, and the energy released is in the form of radiation and kinetic energy. Much of it becomes thermal energy, which is used to heat water and to convert it to high-pressure steam that drives a turbine. A generator converts the turbine's mechanical energy into electricity.

With solar energy, radiation from the sun is converted either to thermal energy or directly into electrical energy. The former is easier to do. Solar collectors, typically blackened metal plates covered with glass, convert the radiation into heat—for example, for hot-water heating. Alternatively, solar radiation can be converted directly into electricity using photovoltaic cells. A small electrical voltage is generated in these cells when light strikes the junction between a metal and a semiconductor (or between two different semiconductors).

Each of these processes uses different energy sources and technologies to convert the energy into electricity. All of these processes have various by-products, emissions, and other residual effects, and it is these various effects that can affect the environment and public health.

Environmental Impacts

The environmental impacts associated with energy use are caused by pollutants and other burdens on the environment that are discharged as part of the many different types of energy conversion processes. There is no escaping the fact that energy use has environmental impacts.

But there is more to this story. Other activities take place both before and after energy conversion into electricity. For example, to generate electricity using coal, a chain of activities must occur:

- exploration for coal
- development and operation of a coal mine
- treatment of mined coal through a process called *beneficiation* to remove rock and debris and to make the coal into a reasonably uniform size
- transportation of the coal by rail, barge, or truck to a power plant
- generation of electricity using the coal
- disposal of solid and liquid wastes and of emissions of airborne pollutants
- transmission of the generated electricity to distributors
- eventually, closing and decommissioning the power plant and the coal mine.

This sequence of activities, necessary for the generation of electricity, is called a *fuel cycle.* The "upstream" activities of the cycle are those that occur before the generation of electric power. The "downstream" activities take place after generation. Each activity is potentially a source of pollutants. Figure 3.1 illustrates the fuel cycle concept.

Table 3.1 lists a wide range of environmental impacts that can result from alternative processes for generating electricity. Many different types of impacts can occur, including occupational hazards, risks to human health from airborne

Figure 3.1. Major activities and emissions of fossil fuel cycle analysis.

Table 3.1. Impacts Associated with Different Electricity-Generation Fuel Cycle Activities

Type of fuel and technology	Occupational hazards	Public health impacts	Environmental impacts
COAL	Accidents and illnesses connected with production of materials needed for plant and coal mine construction.	Injuries and mortality connected with accidents in coal transport.	Loss of land from open pit mining, or mining damages in underground mine areas, including damages to urban infrastructure.
	Accidents arising from mine construction and operation.	Effects of inhalation of pollutants released during production of materials needed for plant and mine construction.	Pollution of water due to liquid effluents from coal mines.
	Occupational pneumoconiosis (black lung disease), silicoses, lung cancer.	Effects of inhalation of coal combustion–derived air pollutants released during power plant operation.	Pollution of water due to solid and liquid wastes from coal-fired power plants.
	Occupational cancer arising from radon exposure.	Somatic and genetic effects attributable to radiological impacts of the coal fuel cycle.	Loss of forests, crops, and animals due to absorption of coal combustion–derived air pollutants released during power plant operation.
	Accidents during the construction of the power plant.	Impacts of liquid and solid waste containing toxic substances.	Global warming due to the CO_2 releases during plant operations, material production, and plant construction.
	Accidents in transportation of coal to the power plant.		Reduced visibility due to haze.
	Accidents in the operation of the power plant.		

(*continues*)

Table 3.1. Continued

Type of fuel and technology	Occupational hazards	Public health impacts	Environmental impacts
NATURAL GAS	Accidents and illnesses connected with production of materials needed for plant, gas field development, and transport lines construction.	Injuries and mortality connected with accidents in gas transport.	Loss of forests, crops, and animals due to absorption of gas combustion-derived air pollutants released during power plant operation.
	Drilling accidents.	Effects of inhalation of pollutants released during production of materials needed for plant construction and gas field development.	Global warming due to the CO_2 releases during plant operation, material production, and plant construction.
	Accidents during the construction of the power plant, gas field development, and transport line construction.	Effects of inhalation of gas combustion–derived air pollutants released during power plant operation.	Reduced visibility due to haze.
	Accidents in transportation of gas to the power plant and gas storage.	Wastes containing toxic substances.	
	Accidents in the operation of the power plant.	Fire and explosions of stored gas.	
OIL	Accidents and illnesses connected with production of materials needed for plant, oil field development, and transport.	Injuries and mortality connected with accidents in oil transport.	Pollution of water due to liquid effluents during oil transport and accidents.
	Drilling accidents.	Effects of inhalation of pollutants released during production of materials needed for plant construction and oil field development.	Pollution of water due to solid and liquid wastes from oil-fired power plants.
	Cancer risks in oil refinery workers.	Effects of inhalation of oil combustion–derived air pollutants released during power plant operation.	Loss of forests, crops, and animals due to absorption of oil combustion–derived air pollutants released during power plant operation.
	Accidents during the construction of the power plant, oil field development, and transport line construction.		

	Accidents in transportation of oil to the power plant and oil storage.	Liquid and solid waste containing toxic substances.	Global warming due to the CO_2 releases during plant operation, material production, and plant construction.
	Accidents in the operation of the power plant.	Fire and explosions of stored oil.	Reduced visibility due to haze.
NUCLEAR	Accidents and illnesses connected with production of materials needed for nuclear power plant and associated facilities.	Effects of inhalation of pollutants released during production of materials needed for plant and mine construction.	Effects of radiation on plants and animals in the case of severe reactor accidents.
			Water heating by waste heat.
	Accidents arising from uranium mine construction and operation.	Somatic and genetic effects of routine and accidental exposure to airborne, waterborne, and foodchain-borne radionuclides from mining, fuel processing, plant operation, and waste management.	Global warming due to the CO_2 releases during material production and plant construction.
	Occupational silicoses.		Nonradioactive releases/impacts from mining.
	Occupational cancer arising from radon exposure.	Nonradiological accidents during materials transport.	
	Accidents during the construction of the nuclear power plant.	Long-term exposure of many generations to very low radiation fields due to long-lived radioactive gases released to the atmosphere.	
	Accidents in transportation of fuel to the nuclear power plant.		
	Accidents in the operation of the nuclear power plant.	Somatic and genetic effects of exposure to radionuclides during power plant operation (including possible accidents and maintenance), during radioactive waste handling and disposal, and during fuel reprocessing.	
	Accidents in waste disposal and fuel reprocessing facility (in the case of a closed fuel cycle).		

(*continues*)

Table 3.1. Continued

Type of fuel and technology	Occupational hazards	Public health impacts	Environmental impacts
HYDRO	Accidents during production of materials and construction of the plant and water reservoir.	Possible relocation of large populations.	Changes in local or regional climate.
		Atmospheric releases due to production of structural materials and plant construction.	Influence of the water reservoir on fishing.
		Risk of dam breaks.	Negative influence on the neighboring land, becoming partly dry and partly wet, with significant changes in function of the water level in the water reservoir.
		Health problems in the coastal waters (growth of vegetation, muddy water, mosquitoes).	Sedimentation of dams, leading to filling up of the area before the dam and accumulation of toxic substances in sediments.
		Displacement of population, tribes.	Global warming due to CO_2 releases during material production and plant construction.
			Loss of forests, land, crops, and historic sites.
			Water management includes also positive sides, e.g., the possibility of control of floods.
			Loss of plant species.
			Loss of animals and their habitats.

BIOMASS	Accidents and illnesses due to intensive fuel growing and harvesting, fuel processing and handling, fuel transportation, fuel storage (hazards of self-ignition), plant operation, and waste handling and transportation.	Public hazards due to inhalation of pollutants emitted during plant construction, plant operation (total suspended particulates, polycyclic aromatic hydrocarbons, formaldehyde, odor), and exposure to biological pathogenes. Hazards to the public due to fuel transport. Global climate impacts due to CO_2 releases during material production and plant construction.	Environmental effects of fuel preparation: (1) odors from stored fuel and (2) land occupied due to monoculture and possible problems with biodiversity.
SOLAR	Production of materials needed for photovoltaic cells (aluminum, silicon, steel). Accidents during material transportation. Plant and backup or energy storage system maintenance. Routine and accidental exposure to toxic toxic chemicals used in device fabrication.	Inhalation of pollutants emitted during material production. Health costs of energy backup or storage systems.	Plant and backup or energy storage system construction. Routine and accidental exposure to toxic chemicals released in device fabrication and in disposal of equipment. Land use.
WIND	Accidents and illnesses due to production of materials needed for wind technology equipment and their transportation, construction of the plant, maintenance, and dismantling.	Inhalation of pollutants emitted during material production.	Global climate impacts due to CO_2 releases during material production and plant construction. Hazards to birds. Land use.

(*continues*)

Table 3.1. Continued

Type of fuel and technology	Occupational hazards	Public health impacts	Environmental impacts
	Accidents during material transportation.	Noise.	
TRANSMISSION LINES	Accidents due to construction and maintenance of transmission lines.	Breakdown of the lines.	Hazards to birds.
		Health effects of electric and electromagnetic fields.	Land use.

Source: Adapted from Russell Lee et al., *Health and Environmental Impacts of Electricity Generation Systems.*

as well as other pollutants, and impacts on ecosystems from discharges of airborne, liquid, and solid pollutants.

Emissions of pollutants can take place in one or more of the activities in the fuel cycle. For example, airborne particulate matter is emitted in the extraction of fuel as well as in processing and transporting coal, oil, gas, biomass, and uranium. Several of the energy conversion technologies are similar in terms of their airborne pollutants. For example, coal, oil, gas, and biomass power plants all emit nitrogen oxides. However, there are also major differences among the technologies. For instance, nuclear power plants emit negligible quantities of air pollutants. In general, the energy conversion processes of different energy sources differ and, as a result, so do their emissions and other burdens on the environment.

For most renewable energy sources, the upstream activities are usually less extensive than those for other energy sources. For example, with wind energy, no fuel needs to be extracted and transported, unlike for electric power from the fossil fuels, where coal, oil, or natural gas is extracted from coal seams or geologic reservoirs. With renewable energy resources as well, the downstream activities tend to be limited—for instance, there are no requirements comparable to the need to store spent radioactive fuel in the nuclear fuel cycle.

However, while renewable energies are generally regarded as having relatively benign environmental impacts, renewable energy fuel cycles can lead to environmental impacts. For example, the wastes of gallium arsenide, a semiconductor used in some photovoltaic cells, are toxic. As another example, the production of concrete, used in dams and other retaining and diversionary structures for producing hydropower, has pollutant emissions. These latter emissions are called *indirect* or *secondary emissions* since they are "once removed" from the fuel cycle activity itself.

The nature of the emissions and burdens also varies considerably among the fuel cycles. In the nuclear fuel cycle, release of radioactive materials is the greatest concern. Relative to other energy sources, the use of fossil fuels leads to considerable amounts of airborne emission: particulate matter, sulfur dioxide, nitrogen oxides, carbon dioxide, and many other pollutants, including volatile organic compounds, carbon monoxide, lead, and other metals. There can also be liquid and solid waste discharges, such as contaminated "muds" from petroleum drilling activities and ash from combusted coal.

Some burdens on the environment are not pollutant discharges but other effects. For example, hydropower development can affect nearby aquatic ecology. The effects of hydro development on salmon in the Pacific Northwest region of the United States are a common public concern. Wind farms can kill birds that fly into the turbine blades. Some power projects might have an adverse aesthetic impact, such as petroleum development in an environmentally pristine area.

Estimating Environmental Impacts

Whatever the reasons for estimating the environmental impacts of energy use, it is important to realize that the types and the magnitude of these impacts can vary considerably, not only among different sources of energy but also among options within each broad category of fuel or technology. Thus, it is important to estimate these impacts in consistent, technically sound ways so that different energy conversion alternatives can be compared consistently. A methodology called the *impact pathway approach* is used to estimate the environmental impacts of energy use.[1] This approach models the physical path of a pollutant, beginning with its source, then its dispersion, and finally its intake by people (for example, through breathing air with pollutants) or by ecosystems. Some studies further estimate the social or economic "value" of these damaging impacts to society.

Estimating the environmental impacts of electric generation has four main steps:

1. Define the specific energy technology under consideration. This means describing the technology and identifying the type of fuel it uses, the power plant's efficiency, and its different discharges. Engineering detail is, in general, unnecessary for this analysis—what matters most are estimates of the different types of discharges and other effects of the technology. Generic data on the types and levels of discharges, such as those from the U.S. Environmental Protection Agency, or, preferably, data specific to the options under consideration should be compiled. The results of this first step of the analysis are estimates of the types and quantities of emissions and other burdens. In some instances, the scope of a study is such that the major emissions are identified for the whole fuel cycle. Figure 3.2 illustrates this approach.

2. Use *dispersion models,* which simulate the spread of a pollutant through the air or water, to estimate the concentrations of different pollutants at different locations. This analysis is used because environmental impacts result from exposure of people and ecosystems (sometimes collectively referred to as *receptors*) to pollutants and other burdens. The concentrations of the pollutants at the locations where the receptors are exposed—not the concentrations at the emissions' sources—determine the type and magnitude of expected impacts. Thus, it is necessary to estimate the dispersion of emissions. Software is available (from the

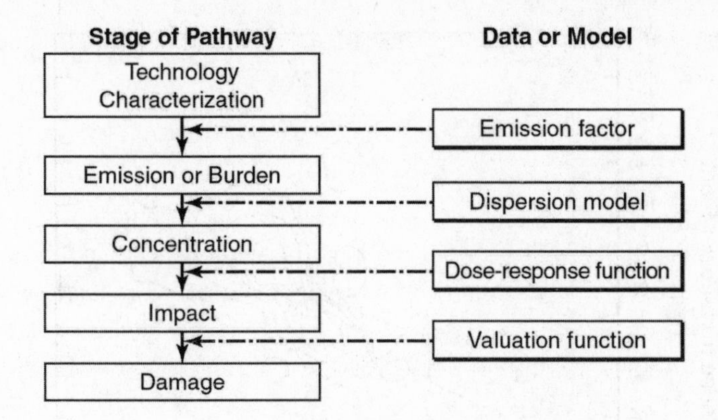

Figure 3.2. The main steps of the impact pathway or damage function approach (produced by Oak Ridge National Laboratory).

U.S. Environmental Protection Agency, for example) for several different dispersion models. Air dispersion models are the most common. Dispersion of pollutants in aquatic systems is modeled as well, though the models are not as well developed. Many pollutants do not undergo any chemical transformation in the atmosphere as they are dispersed. Other factors being equal, the distribution of these pollutants follows a Gaussian (that is, bell-shaped or normal) distribution. Concentrations are greater near the source of the pollution and decrease as distance increases from the source.

Other pollutants—called *secondary pollutants*—undergo chemical transformations. An example is acid precipitation, formed from sulfur dioxide and nitrogen oxides. The formation and dispersion of secondary pollutants is more difficult to model than pollutants that do not undergo chemical transformation, but there are models that provide estimates for the formation and dispersion of these pollutants. The numerical results from these dispersion models are estimates of the changes in the spatial concentrations of each pollutant. Figure 3.3 illustrates some of the output from a model that predicts ozone concentrations; note the spatial distribution of the changes in ozone concentrations, as reflected in the lines that represent locations that have the same change in concentration, measured in this case in parts per billion. The figure shows that the concentrations can follow a complicated pattern, with some areas experiencing decreases in ozone concentrations (that is, ozone scavenging), and other areas experiencing increases (that is, ozone bulges).

3. The spatial distribution of changes in concentrations of pollutants is combined with baseline data on population, agricultural activities, and sensitive ecosystems. These data are overlaid on the distributions of concentrations of

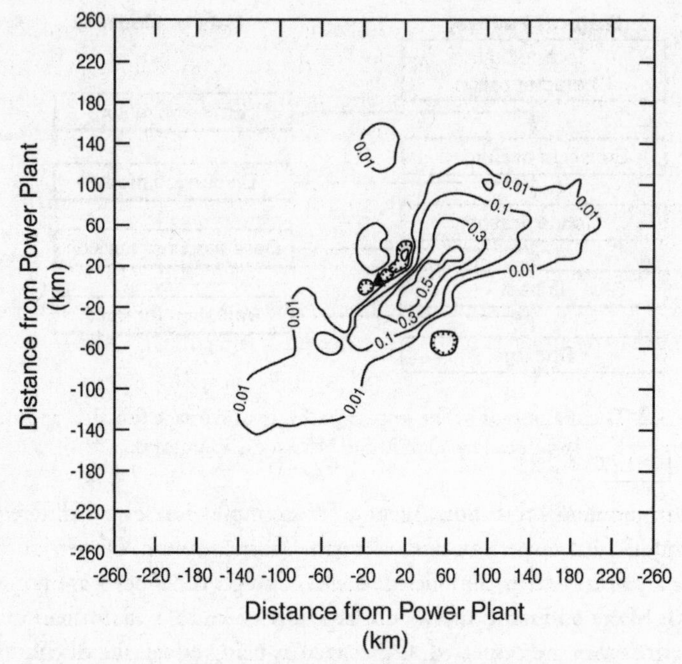

Figure 3.3. Complex atmospheric chemical processes lead to irregular geographic patterns in the concentrations of some pollutants (produced by Oak Ridge National Laboratory).

pollutants, and the resulting estimates of exposure are used in combination with dose-response functions to estimate the expected impacts of these pollutants. Health dose-response functions are equations derived from previous epidemiological studies. Other types of impacts are based on environmental studies. The equations provide estimates of, for example, the expected increased incidence of respiratory illness among the elderly population as a result of exposure to elevated concentrations of a pollutant such as ozone in the troposphere. It is important to realize that these estimates are of the *expected* impact ("expected" meaning the mathematical average or mean value). Environmental impacts are generally probabilistic in nature. Our estimates are inexact. For example, often it is not possible to ascribe a particular case of ill health to increased pollution. Rather, the effect of the increased pollution is to increase the *probability,* or the *expected* number, of illnesses (and in some cases, even the risk of mortality).

4. A fourth step is sometimes used in analysis of environmental impacts. This

step involves changing the estimates of impacts into a different unit of measurement, usually a monetary one—that is, the economic value of these impacts is estimated. The reason for this last step is to facilitate comparison of impacts on the basis of a common unit of measure. Sometimes, regulations or fiscal policies might be instituted to reduce environmental impacts. In these situations, it is common for public policy analysts to assess the costs and benefits of alternative regulations and policies through the use of valuation functions. Economists estimate these equations, which are based on either empirical or experimental evidence of groups of individuals' implicit willingness to pay to avoid an incremental, damaging impact, such as the discomfort associated with a respiratory illness.[2]

Environmental Impacts of Electricity Generation

As we have emphasized throughout this chapter, there are many different environmental impacts associated with electricity generation. The types of impacts depend on the pollutants and other burdens and on the receptors of these pollutants.[3] Of the various possible impacts that Table 3.1 identifies, some are more severe than others. Different types of technologies and engineering processes are used for each type of fuel. Although different power plant configurations and pollution abatement equipment options can be used for each type of fuel, the fuels and their technologies are different. Correspondingly, the discharges and other effects of the various types of power plants are different. In addition, as described earlier in this chapter, associated with the use of any fuel or technology is a fuel cycle. Each activity within a fuel cycle can—and in fact does—itself result in emissions or other burdens on the environment.

Different energy sources and technologies have different types of fuel cycles. For example, with coal fuel cycles, the impact pathways range from the effects of coal dust from mining activities on fibrotic pneumoconiosis (black lung disease) to pavement damages from heavy trucks transporting coal to children cough-day impacts from air pollution (probably due to sulfates formed from sulfur dioxide) to a range of impacts associated with global climate change. Also, as previously emphasized, the severity of these impacts varies with specific conditions, even for a given category of impact pathway. Certain impact pathways, however, tend to be more important. These more severe impact pathways are listed in Table 3.2 and are discussed briefly below. Some impacts, such as biodiversity, are extremely difficult to quantify and are generally best studied on a detailed local level.

Table 3.2. Environmental and Health Impacts of Electricity Generation Using Fossil Fuels

Fossil Fuel Plant—Source	Impact	Relative Severity	Comments
Carbon dioxide emissions	Climate change impacts on health (increased incidence of infectious diseases, increased incidence of hay fever, heat stress during extreme hot weather).	XXX	Global impacts expected to be large. Effects on local and regional residents uncertain.
	Climate change impacts on agriculture and on ecosystems.	XXX	Global impacts expected to be large. Effects on a given region are uncertain. Some agricultural impacts will be positive.
	Climate change impacts on infrastructure, buildings, and land use.	XXX	Global impacts expected to be large. Effects on a given region are uncertain. Low-lying areas at significant risk to sea-level rise.
Sulfates and nitrates from SO_2 and NO_x emissions	Increased mortality and morbidity (illness), principally from respiratory health problems among the very young and elderly population segments.	XXX	Most significant concern of any of the impacts, other than climate change.
Acidic deposition from SO_2 and NO_x emissions	Adverse impacts on terrestrial and aquatic ecosystems (e.g., forests and fish populations, respectively).	X	National Acid Precipitation Assessment Program generally found the impacts to be less than suspected, with the impacts being very site and species specific.
	Impact on freshwater fisheries and recreation.	X	Complex chemistry; no quantitative studies available but not identified as a significant problem.
	Damage to buildings, statues, and other structures.	X	Cumulative damage can be significant to the stonework of old buildings.
Particulate matter from primary particulate emissions	Increased mortality and morbidity (illness), principally from respiratory health problems among the very young and elderly population segments.	XX	Well-researched effects have been identified.
	Soiling of household furnishings.	X	

Fossil Fuel Plant—Source	Impact	Relative Severity	Comments
Ozone formed from NO_x emissions, which undergo chemical transformations in combination with volatile organic compounds, in the presence of sunlight	Increased morbidity (illness), principally from respiratory health problems among the very young and elderly population segments. Also evidence of increased mortality.	XX	Well-researched effects have been identified.
	Reduced crop yields.	X	Experimental results well documented. Impacts are crop specific, and yields are reduced by a small percentage.
	Effects on forests.	X	No quantitative estimates of effects on tree growth, but damage has been identified in some areas of the country.
	Reduced visibility due to haze. Includes value of reduced aesthetic value as well as possible reduction in tourism in scenic areas.	X	Can be very significant for highly valued scenic areas.
Solid wastes and liquid effluents, some containing toxic substances	Pollution of water due to solid and liquid wastes, particularly from coal- and oil-fired power plants.	X	Generally not an issue with existing environmental regulations.
Lead, mercury, arsenic, other discharges	Adverse neurological health effects (reduced IQ among children).	X	Effects very marginal.
	Adverse effects on ecosystems (e.g., fish).		Causative links not well established. Magnitude of emissions generally not a major concern.
Noise from operation of power plant	Noise.	X	Not generally of concern with regulatory control.
Presence of power plant	Reduced visual aesthetics.	X	A "sunk cost" with an existing plant.
Sources other than from the power plant, including fuel extraction, processing, storage, and transportation	Impacts are varied, including road pavement damage from coal trucks, risk of accidents from coal unit trains or from coal and oil trucks, gas pipeline explosions, oil platform accidents, oil tanker accidents and oil spills, coal mining accidents and oil platform accidents.	XX	Impacts might occur outside a given region.

Source: Author.

Fossil Fuel Cycles

In most cases, the two most important impacts of fossil fuel cycles illustrated in Figure 3.4 are effects on the climate from greenhouse gases and effects on health from particulate matter.

GREENHOUSE GASES

Carbon dioxide and other greenhouse gases, such as methane, are emitted when fossil fuels are used to generate electricity. The effects on climate of such an increase in greenhouse gases in the atmosphere are uncertain and are usually considered in several steps. First, the concentration increase due to the emission increase; second, the temperature increase caused by the concentration increase; and third, the impact of the temperature increase. The weight of the evidence in the Intergovernmental Panel on Climate Change and other reports indicates that a doubling of carbon dioxide concentrations (which is almost certain over the next one hundred years) is likely to lead to an average worldwide temperature increase of at least 1°or 2°C.[4] In fact, based on thirty-five different scenarios, projections are in the range of 1.4° to 5.8°C, with the commonly accepted range being 1.5° to 4.5°C.[5] These projections are based on climate models whose mathematical equations simulate processes in the atmosphere, land surface, oceans, vast areas of sea ice, and biosphere based on the laws of physics. The predictions of these models are projections of the changes in average global temperature and water vapor as a result of greenhouse gas emissions over long time periods.[6]

The lower end of the range is serious enough, in terms of the impacts on the environment and societies, but the upper end of the range is likely to be terribly devastating to many regions worldwide. The effects of temperature change on vegetation, agriculture, human health from infectious diseases, and damage from sea level rise are likely to vary widely throughout the world. They will depend importantly on societies' ability to adapt to the changes in climate. But, in any event, the impacts are likely to be significant. The possibility of an increase in the numbers of extreme weather events, such as floods, droughts, and hurricanes, could have dramatically more damaging impacts.

Climate change impacts are of concern for all types of fossil fuel power plants—oil, natural gas, and coal. However, per unit of energy, the impacts are less from natural gas than from coal, because in natural gas half the energy comes from combining the hydrogen component of the molecule and only half produces carbon dioxide. Moreover, the efficiency of modern combined-cycle gas turbines is much higher than that of a coal burner with particulate suppression, thus reducing the specific emissions from natural gas even further.

HEALTH EFFECTS

Epidemiological studies over the last half century were stimulated by the air pollution "incidents" in Pennsylvania in 1947 and in London in 1952. The studies of these and subsequent incidents showed an acute effect of air pollution on health. Death rates increase during and immediately following significant increases in concentrations of air pollutants. However, the most damaging effect is probably the increase in mortality and morbidity rates due to respiratory illnesses from long-term, or *chronic,* exposure. Originally, this connection was suggested in an association of death rates with the concentrations of pollutants in a certain city or area. Such studies are subject to severe error due to confounding factors—for example, if the rate of smoking is different in different areas. But two recent studies—the Six Cities Study and the American Cancer Society Study—are more convincing because they studied individual health outcomes, which take into account individual characteristics such as smoking, although the individual outcomes are still only compared to representative outdoor pollutant concentrations.[7] The strongest correlation of death rates is with the concentration of "fine" particles (less than 2.5 microns in diameter). This finding makes sense since fine particles are readily absorbed in the respiratory tract. For instance, animal studies on guinea pigs have shown that sulfate particles formed from sulfur dioxide are a greater irritant to the lungs than is sulfur dioxide itself and that fine particles are a greater irritant (weight for weight) than are coarse ones.

Particulate matter is emitted from fossil fuel plants. The devices for removing particulate matter at these plants tend to remove heavier, coarse particles. But more importantly, a considerable amount of fine particulate matter is also formed in the power plant plume by chemical processes that change sulfur dioxide and nitrogen oxides to sulfate and nitrate aerosols, respectively. These fine particles are buoyant and can drift with the wind thousands of miles or kilometers before being deposited by gravity or rainfall. Depending on the pollution abatement efficiencies of fossil fuel power plants, the health risks from both primary and secondary particulate matter might be very significant.

Considerable uncertainty still exists about what, if any, chemical attributes of the particles are responsible for the effects on health. It might be the particles themselves, the acidity of the particles, or the chemical nature. Few scientists doubt that sulfates are troublesome, but epidemiological evidence for nitrates has been lacking because air quality monitoring networks have not monitored nitrate concentrations. Since sulfur dioxide can be (and nowadays often is) easily removed at the power plant chimney stack, whereas nitrogen oxides frequently are not, this is an important issue.

Ozone also affects human health and the environment. Ozone is formed from nitrogen oxides and nonmethane organic compounds in the presence of

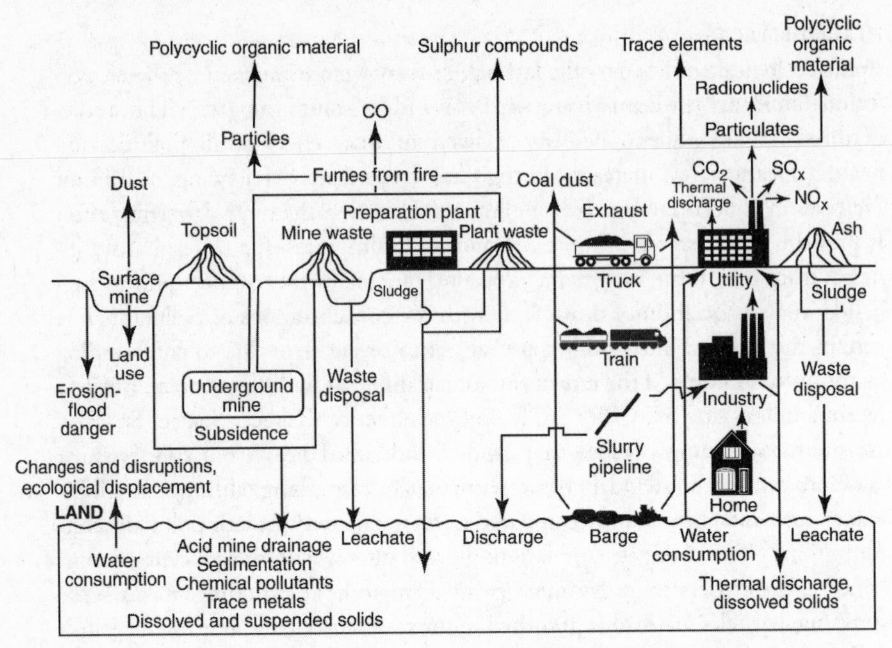

Figure 3.4. Environmental impacts of a fossil fuel cycle.
Source: Oak Ridge National Laboratory and Resources for the Future (1992).

sunlight. High levels of ozone lead to respiratory illnesses and even an increased mortality rate. The dose-response relationship is more uncertain than for particulate matter and could involve a *threshold,* a pollutant concentration below which there is no statistical evidence of health effects. High levels of ozone also cause damage to crops. The effects of nitrogen oxides on ozone concentrations, however, are complicated. Some areas might experience ozone scavenging, leading to a reduction in ozone concentrations. For the most part, however, ozone concentrations increase with increases in nitrogen oxides, particularly in areas with high ratios of nonmethane organic compounds to nitrogen oxides.

Other pollutants, such as lead, arsenic, cadmium, chromium, nickel, and dioxins, also have damaging effects. In general, although they are toxic in higher concentrations, the impact of these pollutants is not as severe because their discharges are usually very small.

In some countries, contaminated liquid discharges might be as severe a problem as airborne pollutants, but their impacts are much more difficult to quantify. There might also be great variability in the discharge of heavy metals, such as lead, mercury, and arsenic. These metals are known to be very damaging in high concentrations. Recently, chronic effects of continued arsenic exposure have been documented in many parts of the world.

Another noteworthy finding is damages from the upstream activities in fuel cycles, such as fuel extraction and transportation. Depending on local and regional mining practices, mining accidents might be very frequent and the risks of fibrotic pneumoconiosis from coal dust might be great. As a final point, distinctive modes of transportation are used to transport different types of fuel. In the case of fossil fuels, there is a possibility of oil spills and natural gas pipeline leaks or even explosions.

Nuclear Fuel Cycles

In the nuclear fuel cycle, severe reactor accidents and waste storage are the greatest concerns. (See Table 3.3 for a list of the environmental and health impacts of the nuclear fuel cycle.) The effect of a severe nuclear accident can be large. But the expected, or average, damages calculated in all modern risk assessments are very small because there is a very small probability of a large release. This probability is small enough that there is no relevant direct historical experience, and so it must be estimated from detailed probabilistic safety analyses. However, public perception often differs from professional judgment. This difference emphasizes a need to understand two different aspects of risk: damage per event and probability. Since it appears that a significant fraction of the public is averse to large risks, the context of the Chernobyl accident is crucial. In that event, almost all of the radioactive gases and a third of the more volatile solid (cesium) were vaporized and globally dispersed. A quarter of the solid material was dispersed from the reactor. In terms of its impacts on the rest of the world, this is as bad an accident as could be expected. While a few early erroneous studies suggested that as many as three hundred thousand persons would develop a fatal cancer, more recent studies suggest the figure might be as low as twenty-five thousand, *assuming a linear dose-response relationship.* If one assumes a threshold or only includes those cancers that might be reliably attributed, the number would be less than one thousand fatal cancers.

Some fuel reprocessing plants discharge low levels of radioactive carbon-14 (C-14). The radiation dose to the general public from this C-14 is a small fraction of natural background (cosmic rays, medical uses of radiation, and radioactivity in the natural environment) and even of the variations in this background. They are well below any applicable radiation standards. But, *if a linear dose response is assumed,* and if the effects are summed across the global population and over several millennia, these releases appreciably increase the long-term health risks over this long time period. If the summation is only over a few hundred years, as some guidelines suggest, or if the economic valuation is discounted so

Table 3.3. Environmental and Health Impacts of Electricity Generation Using Nuclear Fuel

Nuclear Power Plant—Source	Impact	Relative Severity	Comments
Possibility of severe accident	Risk of mortality and morbidity (e.g., cancer) if exposed to high levels of radiation.	X	Highly unlikely event based on probabilistic risk analyses. Public sentiment, other than in communities economically dependent upon these activities, is generally opposed to nuclear energy.
	Damage and contamination of agricultural crops and ecosystems.		
	Destruction and contamination of infrastructure, buildings, soil, and materials.		
Waste disposal	Increased risk of morbidity and mortality from exposure to radionuclides.	X	Waste disposal is controversial. In the U. S., a temporary storage facility is proposed in Utah. Permanent storage is proposed in Nevada. Both states oppose the proposals.
Emissions of radionuclides during normal operation of the power plant—releases can be airborne or waterborne	Very low risk.	X	Extensive probabilistic risk analyses for individual plants have found very low expected risks and corresponding external costs.
Water heating by waste heat	Effects on aquatic life.	X	Can increase stress on fish during summer months but enhance recreational fishing during cold winter months. Thermal impacts are regulated.
Sources other than from the power plant, including uranium fuel extraction, processing, storage, and transportation		X	Occupational risks can be significant.

Source: Author.

that impacts that are expected far into the future are much less important than current or near-term impacts, then these risks are much less significant.

All operators of nuclear power facilities, such as power plants and fuel processing plants, take precautions to prevent theft of fissionable material and its diversion to socially undesirable activities. These precautions add to both capital cost and operating costs. They are not counted as an extra, external cost.

Biomass Fuel Cycles

The greatest health and environmental impacts of the biomass fuel cycle (via combustion or gasification) are those associated with nitrogen oxides and particulate emissions (see Table 3.4). These emissions lead to increased risk of mortality and morbidity as previously discussed. Transportation impacts are also

Table 3.4. Environmental and Health Impacts of Electricity Generation Using Wood and Waste Fuels

Wood or Waste Fuel Plant—Source	Impact	Relative Severity	Comments
Nitrates from NO_x emissions	Increased mortality and morbidity (illness), principally from respiratory health problems among the very young and elderly population segments.	XX	A significant risk.
Acidic deposition from NO_x emissions	Adverse impacts on forests and ecosystems.	X	National Acid Precipitation Assessment Program generally found the impacts to be less than suspected, with the impacts being very site and species specific.
Particulate matter from primary particulate emissions	Increased mortality and morbidity (illness), principally from respiratory health problems among the very young and elderly population segments.	XX	
Ozone formed from NO_x emissions, which undergo chemical transformations in combination with volatile organic compounds, in the presence of sunlight	Increased morbidity (illness), principally from respiratory health problems among the very young and elderly population segments; also evidence of increased mortality.	XX	
	Reduced crop yields.	X	Experimental results have been well documented. Impacts are crop specific.

(*continues*)

Table 3.4. Continued

Wood or Waste Fuel Plant—Source	Impact	Relative Severity	Comments
	Reduced visibility due to haze. Includes value of reduced aesthetic value as well as possible reduction in tourism in scenic areas.	X	Can be very significant for highly valued scenic areas.
Solid wastes and liquid effluents	Pollution of water due to solid and liquid wastes.	X	Generally not an issue with existing environmental regulations.
Sources other than from the power plant, including fuel wood harvesting, waste wood processing, storage, and transportation	Impacts are varied, including road pavement damage from coal trucks and risk of accidents from trucks carrying wood or waste feedstock.	XX	

Source: Author.

significant. These result from trucks that carry the biomass feedstock from the fields to the power plants and in doing so cause air pollution, damage road pavement, and create truck-related traffic accidents.

An advantage of using biomass, on the other hand, is that the net emission of carbon dioxide in the biomass fuel cycle is approximately zero. Carbon dioxide is emitted when tree and plant feedstock is combusted, but these emissions are essentially offset by the carbon dioxide intake of the trees and plants, with carbon being stored in their roots and soil. There are emissions from the diesel equipment used to harvest the trees, but these emissions are small compared to the emissions from fossil fuel power plants. Other benefits of biomass might also include reduced erosion. These impacts, as well as a possible reduction in the number of species (loss of biodiversity) and other ecological impacts, are highly site specific and might lead to either improved conditions or damage.

Hydropower, Wind Power, and Photovoltaic Power

The impacts of other renewable energy projects are highly site specific in both type and magnitude.

HYDROPOWER

Environmental and health impacts of hydropower plants are listed in Table 3.5.

Table 3.5. Environmental and Health Impacts of Electricity Generation Using Hydro Plants

Hydropower Plant—Source	Impact	Relative Severity	Comments
Flow alteration	Impacts on fish population	X	
Changes in sediment load, deposition, or erosion	Impounded water becomes sediment-heavy; discharge water is sediment-deficient; impacts on fish population.	X	
Oxygen-deficient discharged water	Impacts on fish population.	X	
Power plant operation	Turbine-passage mortality of aquatic organisms.	X	
	Gradual, long-term changes in ecology.	X	
Construction of power plant and inundation of nearby area (pertains only to new construction—a "sunk external cost" for existing hydropower projects)	Increased sediment loads, turbidity, or dewatering during construction results in short-term reduction in fish population.	XX	Greatest issues are related to proposed construction of new projects, rather than from continued operation of existing plants. In the latter, the continued operation of the plant does not alter environmental damage incurred from the construction of the power plant and from inundation of nearby areas.
	Blockage or disruption of fish migration or spawning due to changes in water flow reduce the fish population.		
	Creation of impoundment changes water habitat and might reduce indigenous fish species.		
	Impoundment changes land habitat and might reduce population of birds and small mammals.		
	Possible archaeological or cultural losses from inundation.		
	Alteration of habitat or scenic area can reduce visual aesthetics and reduce desirability of the location as a wilderness area; conversely, it can increase the recreational value for water sports and activities.		
	Risk due to dam failure.		

Source: Author.

Large projects that involve the construction of new dams have much greater impacts than those that involve modification of existing dams or small-scale structures that divert water. Accidents during construction, as well as breaches or flooding, may be significant on a local level but are generally small over the operating lifetime of the plant.

WIND POWER

Wind projects might create considerable noise or visual intrusion, which may be particularly bothersome if they are located in areas that are highly populated or highly visited—for example, by tourists. Another concern among environmentalists is that birds are killed by flying into the turbines.

PHOTOVOLTAIC POWER

There are a variety of photovoltaic systems, some of which require the use of toxic substances, such as cadmium and tellurium. The production of these materials leads to air and other discharges. Large-scale photovoltaic projects might also disrupt current land use.

Relative Impacts

It is difficult to compare the severity of different types of impacts. For example, what is the relative severity of greater rates of asthma attacks, to which higher ozone levels might contribute, compared to an estimated reduction in salmon population caused by a hydropower project? Economists, other social scientists, and engineers have tackled such questions in various ways. In general, their assessments of the comparative severity of impacts are based on their estimates of what might be equal tradeoffs among different types of impacts or monetary reimbursement. Examples of the different types of empirical or survey evidence about such tradeoffs are diverse:

- out-of-pocket expenses paid for a recreation trip to a wilderness or scenic area—including all travel costs and taking into account the value of travel time—as a measure of the value of a recreation area
- wage premiums paid for more risky jobs, to reflect a tradeoff between the willingness of individuals to accept a higher-risk situation in return for a higher wage
- estimates of medical expenses, lost wages, or productivity, and the discomfort associated with different illnesses, to indicate the value of different types of morbidity
- the economic value of reduced crop yields (which might be due to elevated exposure to ozone), to measure economic damage

- estimates of reduced real estate values associated with the propensity to avoid living near undesirable and toxic facilities, to reflect the adverse impact of the facility.

Tables 3.2 to 3.5 list results from some of the more prominent studies that have used this literature to implement Step 4 of the impact pathway method described earlier in this chapter. The "relative severity" of the different health and environmental impacts is rated on a scale of X to XXX—the latter being the most severe. This scale is based on a review of the findings in previous studies and reflects the severity of the different types of impacts. An impact with XX is an order of magnitude—or about *ten times*—greater in severity than an impact with X, and an impact with XXX is about two orders of magnitude—or about *one hundred times*—greater in severity than an impact with X. The assessments in the tables reflect a range of both qualitative and quantitative assessments. Of the different types of environmental impacts, the scientific literature supports quantitative estimates of public health impacts better than other types of impacts. Thus, these impacts tend to be quantified more frequently than other types of impacts. This limitation in being able to quantify other impacts does not necessarily imply that these other impacts are insignificant. Rather, it simply reflects limitations in scientific knowledge about their quantitative effects.

Notwithstanding these limitations, the results of previous studies indicate that fossil fuels tend to result in greater impacts on public health and on ecosystems, principally as a result of airborne emissions. Biomass power plants also emit many of these pollutants but have the advantage of very low sulfur dioxide emissions. The biomass fuel cycle also has the advantage that trees and plants use carbon dioxide in photosynthesis. As a result, carbon is sequestered in the roots and in soil, thereby reducing greenhouse concentrations in the atmosphere.

Concerns about proposed new hydropower projects and about certain power plants in pristine natural environments center around the disruption to the existing environment. Concerns about nuclear power focus on safety issues, both of the safe operation of power plants and of the long-term disposal of radioactive waste.

The assessments that Tables 3.2 to 3.5 summarize are generally based on studies that have used the damage function or impact pathway approach discussed earlier in this chapter. This is an important point. The assessments were not based on anecdotes, subjective thinking, or impressions. They were based on sound scientific analysis. The results suggest the following:

- Some energy sources and technologies are usually more environmentally friendly than others. However, we should not make sweeping generalizations

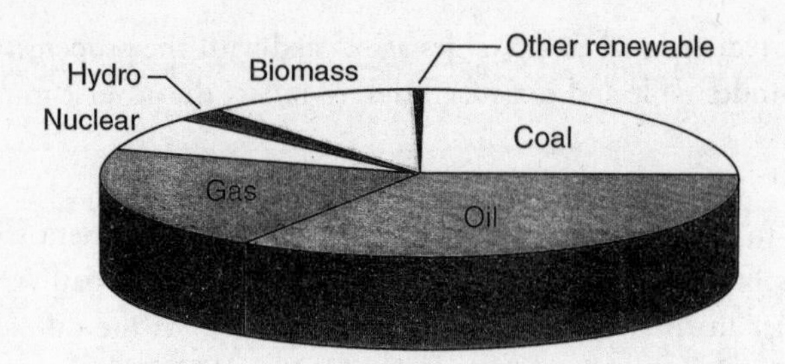

Figure 3.5. Breakdown of percentage use of different energy sources in the world. *Source:* www.iea.org/stats/files/keystats/stats_98.htm.

that some energy sources are "good" or "clean" and that other energy sources are "bad" or "dirty."

- Within different types of technologies, there can be significant variation in the environmental impacts. For example, sulfur content within coal varies, so sulfur dioxide emissions vary. The efficiencies—and thus the amounts of fuel needed and the resulting emissions—of different power plants vary, even among those of the same type. Furthermore, different types of pollution abatement equipment can be installed to reduce emissions.
- Although there are significant differences, none of the sources and technologies is entirely benign. Even renewable energy sources have some undesirable impacts, especially if they are considered on the basis of a full fuel cycle.

Factors Limiting the Use of Environmentally Benign Energy Sources

The most environmentally benign fuels and technologies are not always used. In fact, they comprise a very small fraction of the world's energy use, as reflected in Figure 3.5, which shows that the world's use of energy sources is slanted toward fossil fuels. These fuels tend to have greater environmental impacts than other types of energy sources. Why aren't we using more of the alternative energy sources? There are a number of reasons.

Physical Limitations

The potential for wind power is vast, but it is limited to regions where wind velocities are sufficiently great and sustained. Hydropower can be used only where there is sufficient water flow and the possibility to dam it. Substantial amounts of biomass can be used only where there are agricultural or forestry residues or available good quality land for energy crops. Solar radiation is highest in the tropics and decreases toward the poles. Furthermore, both solar and wind power are intermittent and require storage technology.

Cost

There is an obvious preference for the lowest-cost form of electricity. While good progress is being made in reducing the cost of renewable energies, and wind power is already competitive in areas with high-quality wind, other sources—such as biomass crops and solar power—are relatively expensive except in niche circumstances. For example, photovoltaic electricity at modest levels is competitive in areas distant from a grid.

Quality

In some situations, differences in quality *do* exist between what a conventional energy technology provides to the consumer and what an alternative, more environmentally friendly technology provides. For example, vehicles fueled by methanol have traditionally had problems starting in very cold weather, compared to conventional gasoline-powered vehicles. As another example, electric vehicles have lower acceleration and limited range compared to gasoline-powered vehicles. The intermittent nature of some resources presents additional engineering challenges and costs. Rainfall (necessary for hydropower), wind (necessary to drive wind turbines), and sunshine (necessary for solar power) are not available continuously.

Public Perception

A fourth reason some fuels are not in favor is public perception about their negative impacts or risks. This factor is particularly important in the case of nuclear energy. On the basis of statistical analysis of past performance, as well as probabilistic risk analyses by nuclear experts, nuclear power would appear to be environmentally benign. However, the public's perception of the risks of nuclear power is quite different. The public also has a great aversion to the *nature* of the possible illnesses and deaths associated with these risks. These concerns make the nuclear option very limited in some countries and nonexistent in most.

Cultural Considerations

A fifth reason some energy sources are not as prevalent as they might be in some regions is that cultural or spiritual considerations raise opposition against these options. A prime example of this situation is hydropower. Native Americans frequently object to these projects because they disrupt stream flow, which is of great spiritual importance to Native American culture.

Externalities

A sixth reason relatively environmentally benign energy options are not used more is that decisions about which energy source to use do not completely reflect the full social costs of each option. The social cost includes not only the financial cost but also the *externalities*—that is, the full cost to society. Externalities are the effects on the well-being of individuals and firms that are not taken into account in the markets for the goods and services whose production results in these impacts. For example, emissions of particulate matter from power plants increase the likelihood of respiratory illness, particularly among the elderly and young populations. These impacts have a direct bearing on these populations' well-being, but the impacts are not factored into the market decisions to generate or consume electric power. The existence of externalities means that the market is not taking into account all of the social costs of energy conversion.

Conclusion

Some degree of uncertainty and imprecision always exists in estimates of the health and environmental impacts of energy conversion processes. The uncertainty is simply a reflection of our imperfect knowledge about the dispersion and formation of pollutants, the effects of different pollutants on different plants and animals, and the epidemiological relationships between human health and exposure to different pollutants. Adding to this uncertainty is even greater uncertainty about the *cumulative* effects of different pollutants and of the buildup of the same pollutant from many sources.

In the early sections of this chapter, we discussed how energy conversion is necessary to produce forms of energy that are more useful. We then pointed out that physical and chemical reactions in the conversion process result in discharges of pollutants and other burdens on the environment. Though none of the energy sources and technologies for energy conversion is completely benign with respect to the environment, some are friendlier to the environment than others. We have pointed out some of the reasons why relatively benign energy sources presently produce only a small portion of our energy.

Too often we hear outcries that we have to save the environment, the presumption being that we should do so at any cost. Just as often, we hear that we should let the market decide what and how energy should be produced and consumed, the presumption being that existing environmental regulations provide an adequate safety net for the environment. But the "right" answers are probably not so simple.

Advances in science and engineering *can* make a difference. For example, emissions from coal-fired power plants are much lower now than they were a few decades ago. Federal governments should continue to invest in research and development of more efficient and environmentally friendly technologies.

Nevertheless, science and technology are not the only answer. The global population and its demand for energy are expanding exponentially. Science and technology have done a good job of keeping up with this demand but may not be able to continue at an acceptable environmental cost. The very nature of consumer demand itself may have to change. Sustainability is fundamentally a moral issue, but how to resolve it is still a conundrum. (The moral aspect of this issue is discussed in Chapter 7.) Public policies and related legislation and regulations are part of the answer, especially when social costs and the public good are inadequately addressed by private markets. But the very nature of a democratic society will mean that imperfect, "second-best" solutions will be the norm.

Key Ideas in Chapter 3

- There is no such thing as a completely "clean" energy source with respect to the environment, but some energy conversion technologies are friendlier to the environment than others.
- The two most important impacts of fossil fuel cycles are effects on the climate from greenhouse gases and effects on health from particulate matter.
- Scientific comparisons of the environmental impacts of different energy technologies are indispensable to the public interest in developing energy and environmental policies.
- Relatively benign renewable energy sources presently produce only a small portion of the world's energy.
- Factors working against the use of environmentally benign energy sources include physical limitations, production costs, the quality of the energy service provided, public perceptions of risks, cultural concerns, and limitations of markets to reflect the full social costs of fossil fuel use.
- Advances in science and technology can make a difference but are not the whole answer. Keeping up with world energy demand in the twenty-first century at acceptable environmental cost will require stabilization of consumer demand as well.

Notes

1. During the period from the mid- to late 1990s, a widely accepted methodology emerged. This methodology is called the *damage function approach* or, alternatively, the *impact pathway approach.* The term *damage function* stems from the methodology's use of equations that describe the environmental or health effect that results from a given concentration of a pollutant, which the scientific and economics literatures have previously estimated. The term *impact pathway* comes from the sequential nature of the approach, which follows a path of analysis that estimates the impacts resulting from the discharges of pollutants. The steps in the analysis correspond to the basic steps by which a discharge results in an environmental impact. Much of the appeal of the damage function or impact pathway approach stems from its reliance on the scientific literature rather than on sweeping subjective assessments. Figure 3.2 illustrates the main steps, or stages, in the damage function or impact pathway approach. The numerical results from each step of the analysis are the input for the next step.
2. More detailed descriptions and examples of these methods are given in Oak Ridge National Laboratory and Resources for the Future, *Estimating Externalities of Fuel Cycles,* 8 vols. (Washington, D.C.: McGraw-Hill/Utility Data Institute, 1992–98); European Commission, *ExternE: Externalities of Energy,* 6 vols. (Luxembourg: European Commission Directorate General XII, 1996); RCG/Hagler, Bailly, Inc., and Tellus Institute, *New York State Environmental Externalities Cost Study* (Albany, N.Y.: Empire State Electric Energy Research Corp., 1995); and R. Lee et al., *Health and Environmental Impacts of Electricity Generation Systems: Procedures for Comparative Assessment,* Technical Report Series no. 394 (Vienna: International Atomic Energy Agency, 1999). The important point here is that estimates of environmental impacts should be based on sound scientific analysis, not on anecdotes, emotional appeals, or one-sided opinions.
3. Material in this section is largely based on discussions held during meetings of an International Atomic Energy Agency research program, in which the author participated. A summary of these discussions was written by A. Rabl of École des Mines, France (unpublished).
4. Intergovernmental Panel on Climate Change, *Climate Change 2001: The Scientific Bases,* a report of Working Group I of the Intergovernmental Panel on Climate Change, eds. J. T. Houghton et al. (Cambridge: Cambridge University Press, 2001).
5. Intergovernmental Panel on Climate Change, *Climate Change 2001,* 13.
6. Intergovernmental Panel on Climate Change, *Climate Change 2001,* 48.
7. D. W. Dockery et al., "An Association between Air Pollution and Mortality in Six U.S. Cities," *New England Journal of Medicine* 329:24 (1993): 1753–59; Pope, C. A. III, M. J. Thun, M. M. Namboodiri, D. W. Dockery, J. S. Evans, F. E. Speizer, "Particulate Air Pollution as a Predictor of Mortality in a Prospective Study of U.S. Adults," American Journal of Respiratory Critical Care Medicine 151 (1995): 669–74.

Chapter 4

Culture and Energy Consumption

RICHARD R. WILK

Given our sustainable energy goal, we clearly have problems at both ends of the energy transformation process. This energy process is in the service of meeting human wants and needs. Beyond physiological necessity, what determines these wants and needs and the translation of one into the other? There are surely important biological factors at work here, but the way these factors play out is subject to strong cultural determinants. Social science has wrestled with the issue of human wants and needs since the beginning of the modern era but with limited success.

This chapter is about "consumerism" and the determinants of consumption patterns. We examine the culture of the entire production-consumption phenomenon in both rich and poor countries and illustrate that the many theories on the determinants of consumption patterns are potentially more complementary—even synergistic—than contending. Understanding the cultural, economic, and psychological determinants of consumption is an essential ingredient both in the development of a conservationist ethic that is compatible with human experience and in the formation of an effective long-term energy policy.

—Editors' note

What Is Culture and Why Does It Matter?

Unlike many other species living on the planet, human beings are born with no knowledge of how to survive in their environment and few instincts that would help them survive their first few days, much less their lifetimes. We are relatively unspecialized: we lack sharp claws or thick fur; our eyes, ears, and noses are relatively weak; and we do not run or climb particularly well. Yet, in numbers and sheer geographic spread across the planet, we are extremely successful—entirely because of our unique capacity for culture.[1]

If we make an analogy to computers and liken our bodies and our genetic capabilities to *hardware,* then culture is best thought of as *software.* It includes everything we learn about how to live, the languages we use to communicate, our philosophy, our technologies, and even our ideas about nature, sex, and religion. The key to culture is that it is learned, not encoded in our genes. And it is not learned individually but in social groups, such as families and schools, so communities end up sharing a great deal of cultural knowledge. Culture is both a medium of communication and the substance that is communicated.

Most of the time, we are completely unaware of our own culture—it is like water to a fish, just a part of the environment that we swim through, a seemingly natural part of the world. But while all human beings have culture in the generic sense, as a species we are also divided up into individual *cultures,* which differ widely in their ideas, behaviors, languages, and beliefs. When we compare and contrast those cultures, we find out that many things that we take for granted as normal and natural are actually cultural and, therefore, variable between groups of people.

One good example might be our ideas about comfort. People in the United States tend to believe that a comfortable house is kept heated in winter and cooled in the summer, so there is a relatively constant temperature. Engineers have been doing experiments for many years to try to measure the ideal temperature and humidity ranges for work and home. In other cultures, though, we find very different ideas about comfort and heating. Swedes, for example, find American homes terribly overheated, overcooled, and underventilated. Until very recently, houses in Japan were not heated in winter at all. Japanese people heated the space under a sunken dinner table and trapped the heat with a cloth around the edge of the table. This was not a hardship—people were genuinely more comfortable this way—and in Japan there is still widespread resistance to domestic space heating and cooling. Even in the United States, it took more than thirty years to convince Americans that air conditioning was not going to cause disease.[2]

The key point here is that culture underlies many of our energy-using activities, at many different levels. At the most general level, we can think about the American cultural ideal of the single-family home on its own property and see how this aspect of our culture helps drive urban sprawl, the growth of highway networks, and the amount of fuel we use commuting from home to work in distant cities. Compare this with Denmark or Holland, where even wealthy people are happy to live in high-density apartments or detached homes close to public transportation, where they can also draw home heating from efficient central plants. An equally dramatic cultural effect on energy efficiency is simply the varying size of families and households among different cultures. Because of economies of scale, a large family living together in a house uses energy much more efficiently—on a per capita basis—than a small family. More people are using the same water heater, refrigerator, space heating, and appliances. A comparison of Northern and Southern Europe shows that the smaller households in the north use more energy per person, despite more energy-efficient appliances and houses, than the larger households in the south.[3] The long-term decline in U.S. household size and the ever-increasing numbers of singles living alone are feeding a continuing rise in home energy consumption. Why are we willing, and often happy, to live alone? People from other countries often find this aspect of American culture perplexing and even disturbing.[4]

Perhaps the most important aspect of North American culture that contributes to our standing as the number one energy-consuming country in the world is the general importance of consumption in our lives. Scholars have found various ways to measure *materialism*—the belief that material possessions can bring personal happiness—among different cultures with surveys and questionnaires. By most measures, Americans are highly materialistic, while there is a good deal of variation among other countries both rich and poor. This means that in the United States we tend to focus on material comforts and economic wealth when we think about the quality of our lives. In other cultures, health, family, religion, community, and the state of the natural environment play larger roles in defining the quality of life, so that personal and social well-being is not so directly measured by income or the size of house, car, and TV screen.[5]

Cross-cultural comparison shows that the linkage between increasing wealth, better quality of life, and high levels of energy consumption is weak, for a number of reasons. Some countries seem to do "better" (meaning that their citizens express greater happiness) with less. But in general, richer countries consume a good deal more energy per person than poor countries. There is also a historical trend toward increasing energy use per capita in many different countries. Is this an inevitable development, or does it grow from particular historical circumstances? Finding a solution to the problems of energy consumption in the future depends partially on understanding how we got into our situation in the first place.

The Evolution of Energy Use

Shortly after World War II, the anthropologist Leslie White proposed that human history could be divided into periods on the basis of what kinds of energy were being harnessed by human beings. Beginning with human muscle power, he charted the stages of progress through the steam engine toward nuclear energy. Most importantly, in that optimistic time, White thought that every increase in available energy led to an improvement in human standards of living because it allowed us to produce more goods for each hour of labor. In the American ideal of progress, high productivity leads to more widespread wealth and to more egalitarian and democratic societies. White saw this continuing growth and improvement as an evolutionary law and felt that it would inevitably extend around the world, transforming each society through abundance. Like many other philosophers of abundance at that time, White figured that once energy was cheap and all goods became widely accessible, people would be free from necessity and materialism and would turn to education, music, literature, and other "higher" pleasures while seeking more leisure time instead of working so hard. Many thought we would have a twenty-hour workweek by the end of the twentieth century.

At the time White was writing, few people were thinking about the environmental or social costs of high levels of energy use. Few paused to point out that as more energy became available, the yield in goods and services per unit of energy was in a steep decline. For example, studies of "primitive" farms in Africa found that for every calorie expended by a farmer in planting and tending the crop, the farmer could expect to get about sixty calories of crops. The ratio is almost exactly the reverse in the most "modern" forms of pig raising in the United States, where sixty-five calories of energy are expended to produce a single calorie of pork. Some advanced greenhouse operations require six hundred calories of energy to make one calorie of food. The difference is that the African farmer's calories come from hand labor with a hoe, while most of the calories to raise an American pig come from burning fossil fuels.[6] As long as energy is cheap and abundant and we don't have to worry about the effects of extracting and consuming it on the environment, the loss of efficiency is much less important than the increase in total output.

In a world where energy costs are starting to go up and we are more aware of the environmental costs of consumption, White's ideas about progress seem to be wishful thinking. Instead of sweeping the world in a wave of prosperity, poverty, hunger, and inequality are continuing, and often worsening, global problems. There are more poor and destitute people on the planet today than there were when White was writing in the 1940s, and the gap between rich and poor increased considerably in the late twentieth century, after some years of decline. While we worry about the environmental impacts of high fossil fuel use among the hundreds of millions of inhabitants of the most developed areas, more than a billion people on the planet have little or no access to fossil fuels at all and are still using their own muscles or animal power to wrest their daily living from the earth.

We tend to assume that poverty is the only reason these other people do not live and consume the way we do, but this ignores the role of culture in defining peoples' goals and aspirations. Those who study other cultures have found that, around the world, most people simply do not aspire to the same kind of material wealth that middle-class Americans take for granted. They have other values and other visions of the good life, which may include many items of modern consumer goods but rarely the same ones that Americans value. They may love Hollywood movies and watch *Baywatch* on television, but they value their communities and beliefs and often define happiness and the good life in nonmaterial ways. A high-pressure, career-oriented, mobile life rewarded with a big house, a car, and high-speed Internet access is not everyone's idea of paradise. Even in the United States, many people live relatively simple lives on ranches and farms out of choice, not simply because they lack the resources to move to New York City. At the same time, aspirations to high-consumption lifestyles are widespread across the planet, and no country is untouched by the global consumer market.

So an evolutionary transformation of the kind White imagined is not sweeping the world like an inexorable tide. Despite revolutionary developments in high-speed communications and broader access to travel and media, the world continues to be diverse and unequal. But if it is not some automatic evolutionary force that pushes societies toward higher levels of energy consumption, what is responsible? It turns out that there are instead specific historical and cultural changes that drive consumption in modern society, and we have to place energy consumption in this broader context to understand the nature of the whole problem.

The Origin of Consumer Culture

How is a mass-consumer culture different from other kinds of cultures? People in every society consume goods that are made by other people; there is no group of people on the planet that is completely self-sufficient, so what makes mass-consumer culture different? The key is the amount and type of goods people buy. While early American farmers bought iron tools from the local blacksmith, sugar from Caribbean plantations, coffee and tea from far-away lands, and some fabric from urban mills, the bulk of their daily living still came from their own efforts. So pre–Civil War rural Americans were like most people who came before them; they were only partly self-sufficient and they consumed a variety of manufactured goods, but they were not *mass consumers.* For most historians, the age of mass consumption begins when we see:

- greater volume of mass-manufactured goods for daily use (not just luxuries)
- these goods consumed by a wider percentage of the population (not just by the rich or urban classes)
- broad participation in a *fashion system* of obsolescence and new tastes
- close identification between the person and what that person consumes.

The idea of a fashion system is that many kinds of goods and practices—including clothes, foods, houses, body adornments, conveyances, games, music, and art—become widely consumed in society and that there is constant innovation, so new goods and practices constantly displace old ones. Such a system can thrive only when people are no longer completely identified by their occupations, families, and offices (as in feudal society) but are known instead by the things they own and wear. To some extent, you become what you consume, instead of consuming according to your station in life.

Chandra Mukerji, in *From Graven Images: Patterns of Modern Materialism,* claims that modern consumer culture began in the fifteenth century in Northern and Southern Europe, with a trade in goods among nobles and royalty that included books, relics, jewels, and rare imports of cloth and spices from the Orient. As new groups acquired wealth and sought social status, they became consumers too. Simon Schama, on the other hand, finds the first mass-consumer society in Holland in the late sixteenth and early seventeenth centuries. A new, prosperous middle class began to build comfortable houses and furnish them with paintings and fine furniture while giving heavily to the churches and charities to maintain their respectability and to deflect criticism of their new wealth. Both of these historians think that mass consumption began well before the Industrial Revolution, contrary to the traditional story that new industries allowed consumer culture to grow. The new version says that the Industrial Revolution began *because* there was a new consumer demand for goods among much more of the population.[7]

Many other historians believe that true mass consumer society began much later, encouraged by the new industrial processes that made goods cheaper, by a huge growth in the number of retail shops, by worldwide empires that brought the fruits of hundreds of foreign markets to European shops, and by the invention of advertising and mass marketing in the nineteenth century. It is clear that by the end of the nineteenth century, mass consumption dominated Europe and had a strong foothold in cities around the world, and a good deal of this economy depended directly and indirectly on fossil fuels. Gas and kerosene lighting and coal-powered furnaces, ships, and locomotives used fossil fuels directly. More important in the world economy was *embodied energy,* the energy used to extract and manufacture items such as steel tools and canned beef, which were then shipped around the world. This meant that even people who used few fossil fuels in their own daily work and living ended up using goods that were made through the use of fossil fuels somewhere else, perhaps all the way across the planet. As markets grew and more goods were imported, there was tremendous growth in the energy spent on shipping, trucking, and air freight to move all those goods from one part of the world to consumers in distant places. These trends in consumer culture that began in the nineteenth century have continued to the present.

Today, mass-consumer culture extends across the globe, but its distribution is very uneven. Many people think about the spread of consumer culture as if it were a uniform substance that pours out of a few developed countries and spreads outward, like a wave in a pond, to gradually cover the earth. I often call this the "Sherwin Williams" theory, after that company's advertisement, which shows paint pouring out of a can onto a globe, spreading outward to "cover the world" through the force of gravity. Nothing could be more inaccurate than this image for describing the historical spread of consumer culture.[8]

Cities around the world have been at the center of innovations and development of consumer cultures, but in many parts of the world they are like islands. They are usually surrounded by countryside, where subsistence economies and self-sufficiency prevail. In most countries around the world, the contrast between cities and countryside is now greater than the difference among cities in different countries, in which familiar forms of mass-consumer society can always be found. Cities are the leaders in consumer culture, but they don't always drag the rest of a country along behind them and they are not the only places where consumer culture develops.[9] Some of the earliest places where consumer culture became important were located on frontiers of exploration, on plantations, among miners, pirates, loggers, commercial fishing communities, and traders. Here were uprooted groups of largely young people with cash to spend, who experimented with new forms of consumer goods—for example, new foods. If and when they returned to their rural homes, they often brought new goods and consumer practices with them.

During the past seventy years, mass media and marketing have become more and more important in the spread of consumer culture. In the early twentieth century, some multinational companies experimented with direct marketing in the countryside in the Third World, with theater troupes and performers who traveled from village to village demonstrating products such as soap and kerosene lanterns (this technique is still used in New Guinea and remote parts of Asia and Africa).[10] Hollywood movies, and parallel movie industries that developed quickly in Europe, China, and India, provided powerful images of the "good life," exposing hundreds of millions of people for the first time to the consumer culture of rich and middle-class urban consumers. Since then, many other forms of media have brought images, information, and advertisements to billions.

It is hard to tell what kinds of impact this proliferation of media images has had—surely it had something to do with the increased pace of migration by young people to the cities and with the rising aspirations of people all over the world. But it is quite clear that television and other media do not simply brainwash people into unchecked consumerism. Despite the growth of huge multinational media conglomerates, most media production is still strikingly local in content, and its effects on people are indirect, inconsistent, and highly debated. On the other hand, the many billions of dollars spent every year on advertising and marketing have certainly had a huge effect, turning once-exotic luxuries, such as soft drinks and motorbikes, into expected necessities of life. In seeking to understand the continuing expansion of energy use, this continual growth in needs is a key issue.

The Dynamics of Consumer Society

The British writer Samuel Johnson proposed that human consumption habits set us apart from all other animals on the planet:

> Every beast that strays beside me has the same corporal necessities with myself; he is hungry and crops the grass, he is thirsty and drinks the stream, his thirst and hunger are appeased, he is satisfied and sleeps. . . . I am hungry and thirsty like him, but when thirst or hunger cease, I am not at rest; I am like him pained with want, but am not, like him, satisfied with fulness.[11]

Is it true that human beings are by nature insatiable and incapable of being happy with their lot in life? Most of the major world religions teach the contrary: that true happiness comes only through poverty and simple living, rather than through money and material wealth.[12] Anthropologists have found many cultures around the world in which people are thoroughly uninterested in owning more or having bigger houses, and they do appear to be "satisfied with fulness." This suggests that Johnson is wrong and that, rather than being a natural part of human beings, insatiability is produced by particular historical and cultural situations. Something makes us restless, unsatisfied by what we have, ever seeking new goods that are more fashionable, more comfortable, more satisfying. And we are not just seeking new ways to satisfy the same old needs for food, shelter, and clothing. Instead we are constantly developing new needs; in a process well described by the critic Ivan Illich, wants and desires are gradually

Figure 4.1. The wants-needs spiral.

transformed into needs, luxuries into necessities.[13] The essential characteristic that makes consumer culture so environmentally dangerous is this continual *expansion of needs,* as people come to expect a rising standard of living and measure personal and national progress in terms of increased levels of consumption.

It is tempting to blame constantly rising expectations and the explosion of consumer culture on one particular cause or on some group's moral weakness. For the last two hundred years, writers have been playing a "blame game," pointing their fingers at everything from the bad drinking habits of the poor to a loss of religious faith to the rise of multinational corporations. History shows, however, that the rise of consumer culture has been a continuous process in many parts of the world for centuries, so it is very unlikely that a single cause can be found. Instead, we can make a good argument that many different trends have been pushing consumerism along and helping to expand needs. Biologists use the term *multigenic* to refer to characteristics that are the result of the interaction of multiple genes, such as a person's weight or facial features. By analogy, we can think of consumer culture as being multigenic, in that it has many possible causes, none of which is alone sufficient.

Rather than giving a complete list of all of the causes that have been proposed, presented here are the ones that are backed up by strong arguments and evidence.[14] Each should be seen as a trend that has, at some time, contributed to the expansion of consumer culture.

The breakdown of families and communities and the rise of the individual. In farming cultures, families often act like a small company, in which each person is expected to work his or her hardest and to contribute everything to the common good. When they have some extra money or resources, the family invests in improving the farm and getting more land, not in personal extravagance. So most individuals, especially while they are young, simply have little money of their own to spend. When large numbers of people leave the farm economy and families can no longer claim all of their childrens' earnings, consumer culture becomes possible. Outside the family, many societies have strong institutions that define classes, castes, guilds, or occupations, and people are generally not allowed to rise "above their station" in their consumption. In many societies, *sumptuary laws* define modest and proper dress, diet, and housing for each group. When these rigid categories become more fluid and it becomes possible to change status, consumption often becomes a key element in defining new identities and roles. A more fluid society, in which people find their own individual identities is therefore likely to have much more diverse and flexible forms of consumption.

Status competition and conspicuous consumption. Consumption is the focus of competition in many kinds of society. Even in the isolated highlands of New Guinea, aspiring leaders compete to have the biggest feasts and the most beautiful costumes. But in that culture, only the leaders compete for status—in consumer cultures, just about everyone is engaged in some kind of competition. This broadened social competition, enabled by systems of education and an economic system that allows people to gain great wealth, therefore leads to an expansion of consumer culture. In New Guinea, the rules and the arena of status competition are stable, and leaders compete to provide more pigs and more feathers. In consumer culture, because competition is much broader, with more contestants and a broader audience, novelty is more important. People compete to get the newest items as well as the most and the best. Competition drives a fashion system, which constantly demands new styles and goods.

The need for identity in an insecure world. Many people are familiar with the phrase "you are what you eat." In a consumer culture, in a very real sense, you are what you own, what you wear, what you consume. Psychologists find that Americans, for example, are very much attached to their possessions. They identify themselves, fitting in to an accepted "normal" role in life, through their consumer choices during a lifelong project of choosing cars, neighborhoods, clothing, friends, music, and a host of other possessions. When people lose many of their possessions—whether through catastrophe, illness, or bankruptcy—they often feel completely lost and confused, as if they have lost a part of their identity. This insecurity may be part of the cost of being a mobile, self-sufficient individual with the freedom to make choices. In societies undergoing rapid change or that have an unstable economy or political system, many people will be insecure, driving an insatiable need for more goods.

Comfort and convenience. As societies grow in wealth and harness more energy and materials, it becomes easier and easier to satisfy the basic needs of nutrition, shelter, and physical security. Once these needs are taken care of, people are free to satisfy needs that have previously been put aside, for things like greater comfort, more leisure and entertainment, and self-expression. The growth of consumer culture is a result of greater productivity and more leisure time, which allows people to pursue pleasures they have never before been able to indulge.

Keeping relationships alive. There is no question that Western industrial societies are full of pressures on families and relationships. In these societies, you can buy most of the services—including child care, cooking, and home repair—that were once provided by friends and family. People are mobile, individualistic, and often unwilling to make sacrifices, while jobs and pressures of parenthood make it difficult to keep relationships alive and vital. Shared consumption is one way people manage to maintain relationships that might otherwise weaken or fail. At the daily level, this might mean dining out as a family, giving birthday gifts, or going shopping with a friend. The lifetime events that create bonds between people also require consumption—from a wedding trip to Walt Disney World to a retirement cruise. As relationships become more insecure and fragile, people are driven to intensify the kinds of consumption that build social connections.

Emulation and imitation. One of the ways people learn to be accepted members of any society is to model their actions and ideas on people who are considered role models. They seek to emulate people who are respected and rewarded, and it is so hard to tell this kind of sincere emulation from simple copying or imitation that the two are usually lumped together. After all, they have the same result. In many places, people emulate or imitate respected elders, religious leaders, sports champions, and local politicians. But as European countries expanded and conquered much of the world from the fifteenth to the nineteenth century, new kinds of models were presented for people to follow, particularly through religious conversion and education. All over the world, people were encouraged to emulate Europeans, and soon Hollywood films and other media began to present images of cosmopolitan wealth that also attracted the imaginations of millions around the world. Today, in societies such as the United States, people seek to emulate celebrities and the super-rich, creating an ever-increasing gap between their own lives and those they seek to emulate.

The pressure of marketing. While most people feel that they *choose* to consume, every choice is influenced by what Vance Packard called the "hidden persuaders" of advertising and marketing.[15] While advertising has been with us for a long time, it now absorbs huge amounts of talent, attention, time, and money, all devoted to finding ways to get us to buy more and discover new needs. Most professionals who actually work in marketing admit that individual advertisements are really very weak instruments—it takes a lot of exposure and work to get people to change their habits and tastes, even in something as minor as their choice of toothpaste. Advertisements have no magic power to force people to buy things. Instead, advertising is more like planned seduction, in which sometimes consumers become willing victims. But when people are exposed to thousands of advertisements a week—through radio, television, billboards, newspapers, and even on the walls of schools and on everyday clothing—all those individually subtle influences may add up to a powerful consumerist environment. It begins to feel "natural" to buy something to solve each problem, from a stomach ache to a failed romance.

This is not an exhaustive list of trends. There are many more specific and narrow theories of modern consumerism, but these are the broadest and most general. It is not hard to see that many of these causes are related to each other, and together they suggest that modern, liberal, industrial societies are the most likely to unleash expanding consumerism. But the causes do not just add up: instead, they form systems that feed the growth of new needs.

Cycles of Consumption

Evidence has been accumulating for some time that consumer culture is inherently unstable; it cannot reach some kind of stability because of *feedback.* This means that acts of consumption change circumstances in ways that promote more consumption in an endless cycle. Think about neighbors competing to have the best house. When family A adds a hot tub, family B has to add a hot tub and a sauna, so family A has to put in a pool, and on and on. There is no end to it.

Karl Marx was one of the first to suggest that consumer culture fed on itself through feedback. He said that factory and wage work leads to "alienation of workers from the products of their labors." This means that workers spend their days making things (products of labor), but the things they make don't belong to them. Instead, the employer pays them a wage, which is supposed to substitute for those products. Marx says this leaves every worker with a lingering unhappiness—they have money in their hands, but deep inside they yearn to work for themselves and own what they make. To try to satisfy their unhappiness, they take their money and go shopping. But the things they buy have no real meaning to them, and they cannot fix what is broken, the connection between the workers and the things they make. So the things they buy never really satisfy their deeper needs. They think the problem is that they don't have enough money or enough goods, so they work harder and buy more, acting as if the commodities they buy will have an almost magic power to make them happy and satisfy their needs. But the alienation only gets worse and worse. Eventually, thought Marx, workers would wake up and realize that the only solution was to take over the factories and regain ownership of the products they make.[16] History has, of course, proven Marx's grim romance about consumerism to be wrong, since alienation did not end up driving global revolution and the end of capitalism. But the idea that consumer culture grows through repetitive cycles has recently been revived and given new life.

For example, Juliet Schor has argued that Americans are increasingly trapped in a "work-and-spend" cycle. As we sacrifice more of our free and family time to our careers and ever more insecure jobs, we are driven to get more out of our time. We spend more money on gadgets and services that are supposed to save time and provide convenience, but the things we buy end up requiring our time and attention too, as we set about fixing, replacing, maintaining, and storing all our "stuff." The things we used to do for pleasure and togetherness, such as fixing family meals, are replaced with quick fixes—fast food, gobbled on the run. When we end up feeling more rushed, more pressed for time, and less satisfied with our lives, we buy more new things; to get the money for the new things, we need to work harder still. In the 1960s, when this cycle was really getting under way in middle-class America, it was called simply "the rat race."[17]

Mark and Mimi Nichter are medical anthropologists who have studied eating disorders among children in Arizona. They found that American kids, like their parents, show a great deal of anxiety about their eating, and remarkably large numbers of children are on diets by the time they are eight years old. A majority are dieting and worrying about their weight by the time they turn ten. The paradox is that dieting does not actually lead to thinner kids; instead, dieting and binge eating go together. More concern and anxiety does not lead most people to actually cut down on their average diet. Instead, explain the Nichters, people get in a cycle in which eating becomes both sinful and enjoyable, and afterward they atone and feel guilty by dieting. The guilt eventually fades, and the cycle begins again when people feel that they have suffered enough and deserve a reward.[18]

The Nichters suggest that this sin-guilt cycle is a fundamental part of consumer culture and has become the basis for the main rhythms that mark time in the lives of consumers. Each day is divided into periods of work, when we are supposed to restrain our needs and desires and show discipline, and periods of rest and release, when we are allowed to consume. We get coffee breaks and then go back to work. The day is divided into morning work, a lunch break when we consume, afternoon work, and then we go home and consume. The week itself has the same rhythm of restraint and discipline, followed by the release of the weekend. Years have work time and holiday time, and even our lives have a rhythm in which the hard work and discipline of our working years leads to the relaxation of retirement. This cycle links our consumption to work in a direct way and suggests that the pleasures of consumption are intimately linked to the pain and sacrifice of mostly unrewarding, disciplined work.

While the Nichters focus on the personal level, the entire world economy also goes through cycles that may be driving the growth of consumer culture. Starting in the seventeenth century, European economies began to experience cycles of expansion and contraction, as money markets and speculation led to booms, inevitably followed by busts, when investment capital disappeared, prices dropped, and workers lost their jobs. Over time, the effects of these swings have spread across the globe, and larger proportions of the world's population have been affected by each one. The Internet economy has already demonstrated the same kind of instability. When the economy is expanding, people may be flush with unaccustomed wealth. Growth in the economy spurs optimism and gives people an incentive to buy things they have wished for but have not been able to afford. Those people who do not have new income are still hearing about others who are getting rich. When the inevitable bust comes along, many people have to scale back and accept a drop in their standards of living, but they don't forget what they have given up. Instead, the drop helps build up frustrations, which in turn fuel the expansion of consumption during the next boom. Each cycle acts like a ratchet that moves standards of living ever upward—increasing expectations, increasing consumption.

Overconsumption?

Is there any limit to the amount of goods, materials, and energy that people can consume? Of course, there is only so much food a person can eat and only so many times a day a person can change clothes. But there seems to be no limit to the exotic, rare, and costly ingredients and the elaboration with which food, clothes, and other goods can be prepared. For example, the lips of an exotic reef fish can sell for $20,000 a pair in an exotic Hong Kong restaurant that specializes in endangered species.[19] The super-rich have to hire staff to help them choose, buy, display, and dispose of their acquisitions, but this does not stop them. Every time we think we have seen the most elaborate kind of gown, automobile, or house, something even more extravagant and wasteful comes along. One of the points of conspicuous consumption may be to show that cost does not matter. Drivers of enormous sport utility vehicles may be proving that the cost of fuel does not matter to them, since nothing is too good for them and their families. The vehicles' very extravagant wastefulness is part of their appeal.

When we look at cases of wild extravagance and waste, such as one person owning ten large houses, some of which are never used, or another person buying so many clothes that most are never unpacked, it is easy to define these examples as a form of pathology or sickness that we might call overconsumption. But what about the rest of us? How do we know what is overconsumption and what is reasonable?

We could just average the amount of income and resources in the world and think of the average as the fair share we are all entitled to. But many of the key goods in the world, such as energy and clean, fresh water, are not limited stocks that can be divided up. Standards of living vary a great deal from place to place as well as over time, and definitions of good and bad consumption practices change. At one time, the United States government prohibited the consumption of alcohol in any form, since it was considered dangerous and sinful. Definitions of wealth and poverty have changed continuously, and the United Nations has had great difficulty defining even a minimum level of consumption that meets human "basic needs." Consumption standards and ideas about overconsumption are culturally relative and very changeable.[20] What seems like a basic need in some places—for example, air-conditioning in houses in the Arizona desert—seems like wild extravagance in the equally hot Saharan country of Mali.

Nevertheless, we do need some standards by which to judge both individual and group consumption of energy and resources and to identify particularly wasteful practices. As suggested in other chapters of this book, the concept of energy sustainability provides one way of defining acceptable consumption levels. But in thinking about practical ways to encourage sustainable consumption, issues of "fairness" keep arising—who is to blame, who should pay for solutions, and how much sacrifice should different groups make? When you look closely at the problem of fairness, it turns out that there are actually a number of different ideas about what fairness means that are used by different parties.[21] For example, fairness can mean that everyone gets the same share or that each gets what he or she earns, or that those with the greatest need should get the largest share. The greatest obstacle for the Kyoto Accords in reducing carbon dioxide emissions has been disagreement over whether or not it is "fair" for the rich countries to bear the burden of reduction.

The problem is that we can argue forever about what is fair, while natural systems suffer irreversible damage and vital resources that will be needed by future generations are depleted. With environmental catastrophes of different kinds predicted for the future, perhaps we can settle first for some agreements about simply indefensible kinds of consumption that should be prohibited. In other words, start with simple and obvious cases of overconsumption: wildly inefficient houses, fuel-hog vehicles, antique manufacturing processes. Then we can work on consumption that is more ambiguous and defensible. But what kind of action, if any, can successfully regulate consumption?

Seeking Solutions

There is no doubt that culture is a major obstacle to change in energy use. The ethical system of modern capitalism is the absolute sovereignty and freedom to consume. If you can afford it, you can have it, unless the state has some very strong overriding interest, such as public health or emergency shortages. This ethos is at its strongest in the United States, and Americans tend to think that governments should place very few restrictions on what people consume.[22] Europeans tend to allow their governments to take a much more active role in deciding what kinds of consumption are personally or socially dangerous or which kinds are against the common good. For this reason, they are willing to accept taxes on luxury goods, regulations on energy efficiency in houses and appliances, higher energy taxes, and a host of regulations on waste and recycling that are quite unknown, and would be unacceptable, in the United States. The result is that several European countries have made significant progress in cutting their per capita and total energy use, dramatically increasing efficiency, shifting to renewable sources of energy, and reducing their emissions of greenhouse gasses. But even in Europe, the idea of consumer sovereignty—that we all have a right to spend our own money on whatever we please—has become a key democratic value, superseding the values of civic duty and responsibility to the nation.

Does this mean that personal freedom to consume is so deeply ingrained in modern consumer culture that there is no hope of effective public policies to reduce the consumption of energy and energy-intensive materials and goods? I don't think there is any support for such an extreme position. Even in this most permissive consumer society, Americans continue to recognize that the state and community have legitimate rights to restrict the consumption of individuals when it threatens self-destruction and the common good. These rules range from prohibitions on illegal drugs to building codes and environmental and workplace regulations to tariff barriers, selective taxation, and price controls. While the local superstore presents a vision of complete freedom of choice, behind each good there is a complex network of rules and regulations that are intended to protect the common good (though rarely with complete success). Every society is built on some recognition that the common good can take precedence over individual desires. From this perspective, all that is needed is a stronger political will to reduce wasteful consumption and increase energy efficiency. In the United States, however, industry lobbies have proven extremely powerful political opponents of efficiency standards and regulation. But culture also plays a part. The individualism and independence of North Americans, their suspicion of government, and their intense involvement in consumer culture all make it difficult for the public to support policies aimed at reducing consumption. The administrative culture in the United States also tends to value "hard" technical fixes for problems over "soft" social and cultural approaches that rely on changing behavior. The current Bush administration's energy policy is clearly driven by the idea that technology and the market will provide for the future, rather than any kind of change in consumption and lifestyle.

Unfortunately, as important as culture is to understanding our present lifestyles, our standards of living, and the future trends in our energy consumption and waste production, we often treat energy problems as technical, economic, or political issues instead of cultural ones. Part of the problem is that culture is so pervasive, so much the water we swim in, that it is hard to perceive and is especially difficult to measure and quantify.

Individual and Community Action

Social scientists have been working steadily at understanding the way culture affects everyday routines in ways that directly affect energy consumption. Willett Kempton's work has stressed the importance of giving information back to consumers about their energy use so that they can better understand what they are doing to use so much energy. Small changes in the way electric bills are printed, for example, can lead to a very quick reduction in the amount of electricity people use.[23] Opinion surveys tell us that Americans are worried about global warming and energy issues, and they say they are willing to pay more taxes and fees to solve these problems, but so far these good intentions have not translated into much action. We still need a great deal of research to find out why beliefs and knowledge about energy use and conservation are so rarely acted upon, what kinds of behavior are most wasteful, and how conservation measures can be marketed more effectively. This research will have to extend to every culture in which sustainable energy use is a goal.

In the middle part of the twentieth century, the anthropologist Margaret Mead argued that no single approach to change would work in every society. Instead, she called for a "culturally appropriate" approach to solving problems, which would take into account the unique characteristics of each group of people.[24] Her arguments are still convincing half a century later, and their implications are important for thinking about sustainable solutions to energy problems today. They imply that no single set of solutions—technological, legislative, or cultural—is going to work in every society. On the contrary, we should expect great diversity in the ways different countries and cultures deal with sustainability problems.

How can we expect things to change in the countries that presently consume so much more than their share of global resources? In the United States, many important social changes begin not with government initiatives or university research but in small-scale and very local social movements that start when people meet and talk with one another about their common concerns and interests. Beginning in the late 1980s, a very loosely structured movement began to form around the ideas of encouraging sustainability and building local self-sufficiency. These "simple living" or "voluntary simplicity" groups and discussion circles often focus on helping members to slow down and simplify their lives, consume less, and have more free "quality" time. This may mean working fewer hours or finding a more flexible and less demanding job while developing crafts, gardening, or some other local form of income. They often use a twelve-step program modeled on Alcoholics Anonymous (as presented in the book *Your Money or Your Life*) to break the work-and-spend cycle by helping people become more aware of the daily decisions that lead to unreasonable spending. Many of the groups also seek to educate local communities about the problems of consumerism and world trade, and they may promote local currencies, farmers markets, or other alternative approaches to the globalized consumer economy. Some estimate that more than a million people are involved in this movement.[25]

Conclusion

Perhaps the simple living groups will continue to grow and will link with other forms of environmental activism at a grassroots level to change people's thinking about consumer culture. But we should be very cautious in our optimism, given the extraordinary persistence and power of consumer culture to keep on creating new needs, expanding the links between producers and consumers across the globe. It is worth remembering that in the United States, every increase in energy efficiency over the past twenty years has been eaten up by higher levels of consumption. When the engines in cars become more fuel efficient, people buy bigger cars and drive more. This is the cycle that has to be broken if we are to see movement in the direction of sustainable energy use. Will it take another energy crisis, or a climate change disaster, to make people take the problem seriously? Whatever the future, we can be sure that there will have to be many different solutions to energy problems, each within a local historical, economic, and cultural context. Sustainable energy may be a common goal, but there must be many avenues leading in the same general direction.[26]

Key Ideas in Chapter 4

- Culture underlies human energy use at many different levels. It drives consumption in modern society and is a major obstacle to change in patterns of energy use.
- Energy consumption must be placed in a broad cultural context in order to understand the nature of the sustainable energy problem.
- Culture is not encoded in our genes but is learned in social groups—families, schools, and communities.
- In seeking to understand the continuing worldwide expansion of energy use, a central problem is the continual growth in perceived needs—the conversion of wants into needs.
- The rise of consumer culture has been a continual process in many parts of the world for centuries, so it is unlikely that a single cause can be found or that a single approach to change will work in every society. There will be many different solutions to energy problems, each within a different historical, economic, political, and cultural context.

Notes

1. See, for example, Edward T. Hall, *The Invisible Dimension* (Garden City, N.J.: Anchor Books, 1969), for a basic introduction to the concept of culture.
2. For a discussion of cross-cultural ideas of comfort, see Rita Erickson, *Paper or Plastic?* (Westport, Conn.: Praeger, 1997); and Gail Cooper, *Air-Conditioning America* (Baltimore: Johns Hopkins University Press, 1998), on the struggles to define comfort and get Americans to accept air-conditioning.

3. Jan Klaas Noorman and Ton Schoot Uiterkamp, *Green Households? Domestic Consumers, Environment, and Sustainability* (London: Earthscan, 1998).
4. See, for example, Robert Putnam, *Bowling Alone: The Collapse and Revival of American Community* (New York: Simon & Schuster, 2000).
5. On materialism, see Guliz Ger and Russell Belk, "Cross-Cultural Differences in Materialism," *Journal of Economic Psychology* 17 (1996): 55–77. For quality of life issues, see M. Nusbaum and A. Sen, *The Quality of Life* (Oxford: Oxford University Press, 1993); and Michael Argyle, "Causes and Correlates of Happiness," in Daniel Kahneman, Ed Diener, and Norbert Schwartz, eds., *Well-Being: The Foundations of Hedonic Psychology* (New York: Russell Sage Foundation, 1999), 353–73.
6. Leslie White, *The Science of Culture: A Study of Man and Civilization* (New York: Farrar, Straus & Giroux, 1969). For the opposite point of view, see Robert Netting, *Smallholders, Householders* (Stanford, Calif.: Stanford University Press, 1993).
7. Chandra Mukerji, *From Graven Images: Patterns of Modern Materialism* (New York: Columbia University Press, 1983); Simon Schama, *The Embarrassment of Riches: An Interpretation of Dutch Culture in the Golden Age* (Berkeley: University of California Press, 1988).
8. Daniel Korten, *When Corporations Rule the World* (West Hartford, Conn.: Kumarian, 1995); Richard Wilk, "Emulation, Imitation, and Global Consumerism," *Organization & Environment* 11:3 (1998): 314–33.
9. Saskia Sassen, *Globalization and Its Discontents* (New York: New Press, 1998).
10. One example is discussed by Timothy Burke, *Lifebuoy Men, Lux Women* (Durham, N.C.: Duke University Press, 1997). There is also an excellent film about New Guinea called *Advertising Missionaries* (New York: First Run/Icarus Films, 1996).
11. Samuel Johnson, *The History of Rasselas, Prince of Abissinia,* ed. Geoffrey Tillotson and Brian Jenkins (London: Oxford University Press, 1971).
12. Russell Belk, "Worldly Possessions: Issues and Criticisms," *Advances in Consumer Research* 10 (1983): 514–19.
13. Ivan Illich, *Toward a History of Needs* (New York: Pantheon, 1977).
14. In this discussion, I do not give individual references for each proposition, each of which can be found in numerous individual sources. A reader looking for original sources on each could do no better than to consult Neva Goodwin, Frank Ackerman, and David Kiron, eds., *The Consumer Society* (Washington, D.C.: Island Press, 1997). For the American case, an excellent place to start is Richard W. Fox and T. J. Jackson Lears, eds., *The Culture of Consumption: Critical Essays in American History, 1880–1980* (New York: Pantheon, 1983).
15. Vance Packard, *The Hidden Persuaders.* (New York: Pocket Books, 1980).
16. A good guide to Marx's philosophy of the individual can be found in John Elster, "Self-Realization in Work and Politics: The Marxist Conception of the Good Life," *Social Philosophy and Policy* 3:2 (Spring 1986): 97–126. For the original, see David McLellan, ed., *Marx's Grundrisse.* (London: Macmillan, 1980).
17. Juliet Schor, *The Overspent American* (New York: Basic, 1998), for trenchant critiques of the rat race of American middle-class life.
18. Mark Nichter and Mimi Nichter, "Hype and Weight," *Medical Anthropology* 13 (1991): 249–84. For a similar argument, see Judith Williamson, *Consuming Passions*

(London: Marion Boyars, 1986); and also Gary Cross, *Time and Money* (London: Routledge, 1993).

19. Carl Safina, *Song for the Blue Ocean* (New York: Henry Holt, 1997).
20. Daniel Horowitz, *The Morality of Spending* (Baltimore: Johns Hopkins University Press, 1988).
21. The different folk ideas of fairness are discussed in George Lakoff, *Moral Politics* (Chicago: University of Chicago Press, 1996).
22. Gary Cross argues that consumerism has become the same thing as democracy to most Americans, replacing other forms of citizenship. His historical analysis shows that American consumerism is a relatively recent phenomenon, and it was not an inevitable development. See his *An All-Consuming Century* (New York: Columbia University Press, 2000).
23. See W. Kempton, J. Darley, and P. Stern, "Psychology and Energy Conservation," *American Psychologist* 47:10 (1992): 1213–23.
24. The enterprise envisioned by Mead eventually spawned such enterprises as "appropriate technology" and "applied anthropology." Margaret Mead ed., *Cultural Patterns and Technical Change,* Tensions and Technology Series (New York: New American Library, 1955).
25. Joe Dominguez and Vicki Robin, *Your Money or Your Life: Transforming Your Relationship with Money and Achieving Financial Independence* (New York: Penguin, 1999); and Amitai Etzioni, "Voluntary Simplicity: Characterization, Select Psychological Implications, and Societal Consequences," *Journal of Economic Psychology* 19 (1998): 619–43.
26. See Friends of the Earth Netherlands, *Sustainable Consumption: A Global Perspective* (Amsterdam: Friends of the Earth Netherlands, 1998).

Chapter 5

Energy Policy: The Problem of Public Perception

RANDALL BAKER

Policy, like culture, is a central guiding hand that expresses our collective values and shapes our future. Unfortunately, there are many constraints in the "policy system" of a democratic society that work against a sustainable energy policy, such as its essentially short-term political nature, the national basis of policy making in a global environment, and the unequal voices of interested parties. Furthermore, people have a tendency to react—in terms of policy—to crisis more often than to long-term, evolving situations. What evidence do we have that the public today believes there is an "energy crisis," or that there will be? Moreover, it is easier to establish a momentum of concern when people think that something is being done to them rather than that they are doing something to themselves. For example, consider the public concern over aircraft or nuclear power safety versus the much higher risks people take behind the wheels of their own cars. These are major issues in formulating a sustainable energy policy within the context of complex national and international political and economic systems.

How does everything we know about the laws of nature (including human nature), the finiteness of energy resources, the environmental impacts of energy use, and the steady growth in world population and energy demand play out in the real world—the world of politics and economics? In this chapter, we explore the problem of formulating and implementing policy in a democratic society. We address public concerns translated into policy implementation from the political perspective. Diverse economic and political interests and a tendency to concentrate on policies with near-term impacts are characteristic of, and impose limits on, policy making in a democratic society.

—Editors' note

What is policy? In a democracy, at least, it should be some expression of the prevailing values, the desired futures, and the quality of life that the voting public feel ought to predominate. Policy is normally thought of as positive, with the law being reserved as a sanction for those who knowingly and willfully compromise the standards and mores that the policy embodies. In a functioning democracy, we assume that people have access to the full range of information relevant to their concerns, that they have access to the governance system that effects change in their lives, and that they live under the protection of a system that prevents the abuse of power. So, policy should reflect the concerns of the majority and their changing values.

Thus, when Emile Zola and Charles Dickens were describing some of the worst environmental conditions in the history of mankind, there was no Green Party or environmental movement. The reason for this was the absence, as far as most people—especially those who suffered the worst consequences of pollution—were concerned, of the instruments of participatory democracy. Change, when it came, followed enfranchisement, access to information, and participatory government, not technological advances in environmental management and pollution control.

Generally speaking, in a participatory democracy, the assumption is that policy begins with a widespread concern embodying a perceived need for change. This is sometimes referred to as the *policy agenda.* If the concern is sufficiently widespread, then it transmits itself to the elected representatives, whose job it is to translate a broad consensus of concern into a policy (a positive instrument for change in society) and the accompanying regulations, rules, and laws, which set the standards and establish the sanctions for those who would willfully disregard the "will of the people." This is something of the perfect world so beloved of economic theorists. In practice, there are powerful lobbies, political action committees, political reelection contributions, and other possibilities to apply "undue" pressure. However, in the United States—given the Freedom of Information Act, the openness of the system to lobbying by all parties, the investigative nature of the press, the highly connected nature of the population to the media, and *the unusual ability of the citizen to sue the government for punitive damages*—the policy process is very open. This is not always the case for all developed democracies. However, policy makers have generally been sincere about proposing policy—and following through with it—because in the United States they will be watched and taken to task if a policy is hollow, unenforceable, or window dressing. This was illustrated well by the furor over unfunded mandates by means of which Washington was imposing unenforceable legislation on the states without providing the means to make it work in practice. This was stopped in the congressional shakeup of 1994–95.

It is one of the unwritten rules of policy in most democracies that nothing so fosters the popular perception of a need for change as does a *crisis.* This is a variant of Dr. Johnson's famous observation: "Depend on it, Sir, when a man knows he is to be hanged in a fortnight, it concentrates his mind wonderfully." It could be argued, for instance, that it took the expressive literary genius and scientific wisdom of Rachel Carson to reach a wide audience, but it took the sense of immediate alarm in her message to make them sit up and act, and telephone, and lobby. The title *Silent Spring* was a wonderful evocation of a naturalist's apocalypse. But, the crisis is normally just the flash point, and there also is a need for an underlying nagging concern that something is deeply, and disturbingly, wrong so that this somewhat unfocused psychosis can suddenly be given direction and momentum in the arena of policy making. In terms of the environmental policy awakening of the 1960s and 1970s, there was the anxiety about "the bomb"; the realization through pictures from space that the world really was this fragile and beautiful sphere of very limited dimensions; the runaway pollution of rivers; and so forth.

If we look at the energy question in the last moments of the twentieth century and now the opening years of the new century, then there are some problems about mobilizing significant public concern in any context of policy formulation or change. In terms of a crisis, it is instructive to ask the simple question: how would we know if we had an energy crisis? Since energy is a commodity for which we pay, like any other, and it is supplied and delivered like any other, we can assume that there would be certain signals of a crisis in the making that would alert the public to this condition. Among these should be the following identifiable, though interrelated, elements: supply, price, substitution, and rationing.

Supply

If there is a crisis concerning the availability of a commodity, then the first sign of this should be that the commodity becomes scarcer. Consider, for instance, the recent, sudden drama of rolling blackouts in California. That can be, and has been, portrayed as an energy "shortage." If the lights don't come on, then this seems like a reasonable designation for what is happening. However, what has happened in California continuing through the summer of 2001 has nothing to do with the physical exhaustion, unsuitability, or inaccessibility of *energy* as a natural resource. The "shortage" has everything to do with a badly handled deregulation process and a failure to invest in energy-generating capacity. No one to date has uttered the explanation, "It has run out." The crisis is of political manufacture: bad planning, ineptitude, conniving—take your pick. But no one among the California voters imagines for a moment that the supply has suddenly diminished.

Manifestly, in the eyes of the general public, there is absolutely *no* evidence of a current energy shortage arising from depletion of supplies. Only once has the American—or indeed the European—public been confronted with "Sorry—No Gas" signs since the Second World War, and that was as a result of the brief embargo that followed the Arab-Israel conflict of 1973–74 that took the Western world totally by surprise. But the lines and "No Gas" signs were then, and are still now, attributed variously to strategic exposure (no quickly available alternatives), market manipulation (a small number of countries strongly dominating supply and distribution), political chicanery, an element of cartelism, and a thoroughgoing lack of preparedness (through, for instance, stockpiling) on the part of the rich democracies to face the prospect of what was quite clearly seen at the time as "economic blackmail" resulting from being so dependent on others for something so fundamental to life as we—in the West—know it. In other words, it had nothing to do with a crisis of natural resources or the finite availability of energy from "traditional" fossil fuel sources—it had everything to do with international politics:

> The phenomenon from 1973 on was essentially an *oil* crisis, not an energy crisis. The OPEC embargo, and the sixfold price jump panicked motorists, laid bare America's dependence on foreign supplies, and drained billions of dollars out of the country, setting off an economic recession. Yet the shortage of oil had only a limited effect on other fuels. In spite of all its efforts, the FEA could induce only *one* electric generating plant in the entire country to switch to coal. Total fuel consumption dipped slightly for 1 year, and then resumed its climb upward.[1]

The solutions put in place responded to the strategic implications of a "political crisis" and were to beef up the military presence in those "unstable" places that supply the West's oil; to go to war to protect those resources against those who would interfere with, or try to manipulate, them; to build up substantial strategic supplies at home to buffer such shocks; to develop resources in more "politically stable" parts of the world (for example, Alaska and the North Sea); and to encourage dealing with non-OPEC members. In none of this is there any recognition, in policy terms, that there is an energy crisis per se. In the daily round of things, the energy is always there—and all that matters is securing access to it. In the recent (2000) public protests in Europe, there have been pump closures, "No Gas" signs, frayed tempers, and so forth, but—once more—this is a demonstration of public anger against the *level of energy taxation* by the governments concerned, not an attempt to draw attention to long-term supply considerations.

Price

Normally, when things are in short supply or threatened—and what else would an energy crisis be?—then we would expect the price for this scarce commodity to rise. It is true that a gallon of gas costs more, in dollars and cents, today than it did in grandfather's day—and indeed has seen some really steep price increases in 1999–2000 and in early 2001. But in real terms—adjusted for what we earn—gasoline has still never been cheaper in the United States. Indeed, in 1995, gasoline prices in the United States were lower than at any time in the seventy-seven years they have been recorded. If we compare the price of gasoline in the United States, for instance, to the minimum wage, then we have to work less time today to earn a gallon of gas than ever before (see Figure 5.1). Human energy is, it appears, consistently revalued upward in comparison to fossil energy—even though a gallon of gasoline has about ninety-two times the energy value per hour of human effort. So, in real terms, the price of fossil fuels has actually been *going down* as far as U.S. consumers are concerned (see Figure 5.2).

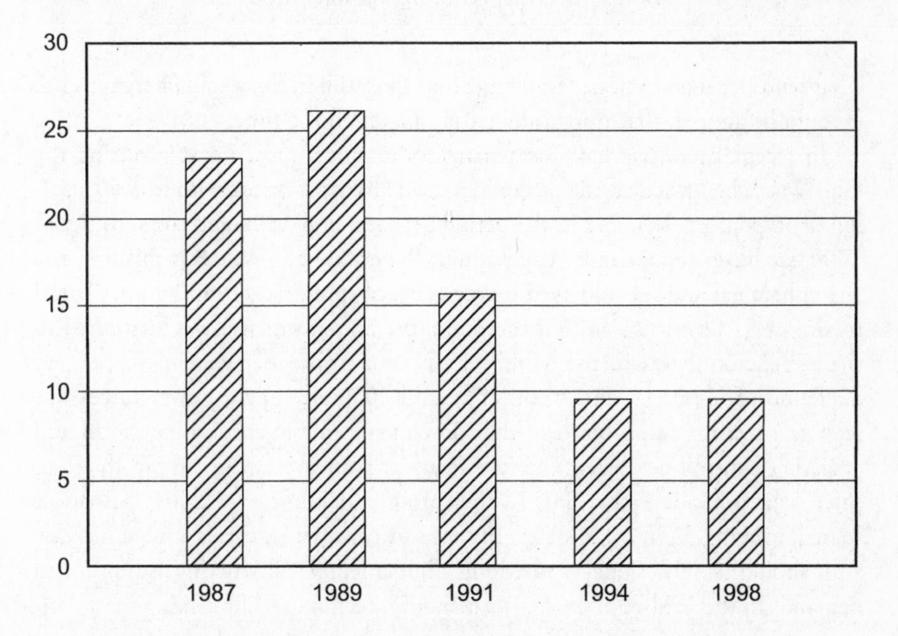

Figure 5.1. Minutes worked at minimum wages (in the United States) to purchase one gallon of gasoline, 1987–98.

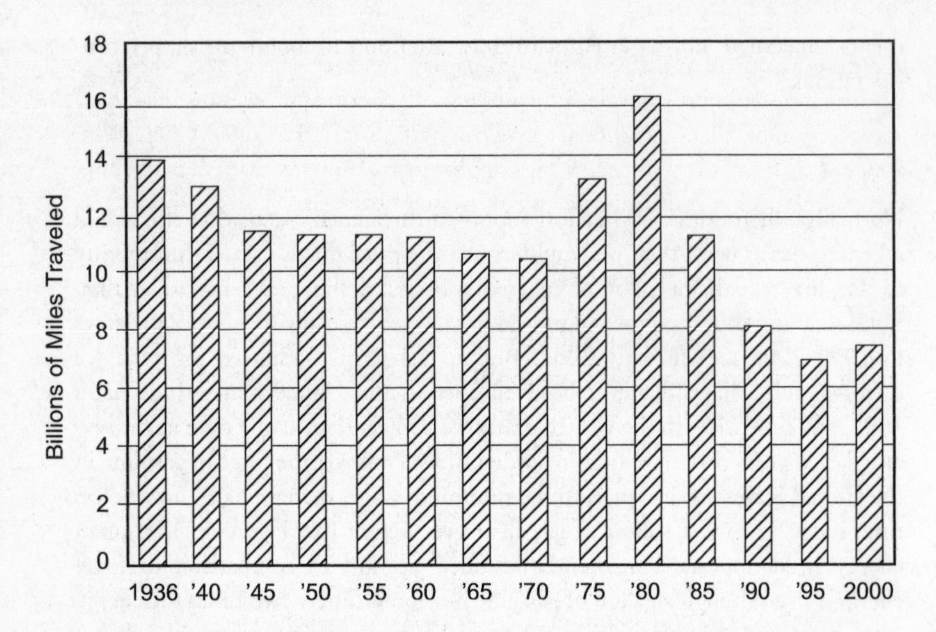

Figure 5.2. Consumer cost of motor gasoline per mile traveled, 1936–2000.
Source: American Petroleum Institute, 1997.

That sends, at the very least, a message that all is well in the world of energy and, it could be argued, that things are getting better all the time.

In recent times, we have seen increases in price, most notably during the Gulf War, when prices at the pump increased almost at once by up to $.50 a gallon (a 50 percent rise), and in the period between mid-1999 and the summer of 2000, we have seen a similar rise without the excuse of a war. But this first rise was ephemeral and was followed by the crash of spot prices from $40 per barrel to almost $10 by 1995–96. Furthermore, the 50 percent rise was attributed to the exigencies of war in the Middle East—not to broader systemic shortages, depletion of resources, and so on. The mid-2000 attempt to lower prices was seen as a confrontation between the "developed" energy-consuming world and a market-controlling cartel, and the focus was on increasing supply to lower price, which Saudi Arabia did. In the spring and summer of 2001, gas prices soared, topping $2 in some states. In terms of percentage, this is a huge increase and should surely signal something fundamental obstructing supply and demand. In the Midwest, the best explanation (echoing California's energy "crisis") was that there was insufficient refining capacity to meet the high environmental standards for air quality. Again, this has nothing whatsoever to do with the finite resources available to us from nature. In other words, price variations are disconnected in the mind of the public from concerns of "wasting resources" or depletion of stocks. Cautionary tales, in 2000, from OPEC about its possible physical inability to go on increasing supplies were countered with tales of "huge new finds" in the Caspian Basin. What is one to believe?

Substitution

When things are in short supply and their price has risen, we normally expect to see substitutes appear fairly rapidly on the market. So, as fossil fuel "marches on toward extinction," as all nonrenewable resources must, and as they form such a tremendously important underpinning of life as we in the "developed" world know it, it may seem curious that there is not more eager attention given to the development of "alternative," renewable energy resources under the impetus of public policy. There was a dramatic refocusing under the Carter administration of interest in these alternatives after the OPEC embargo, but the story since then has been one of diminishing official commitment and resources. The California energy shortage has been seen as the reflection of a need to build more "generating capacity," not to move into alternative energy. Wind power is now cost competitive but supplies only a small fraction of our needs and is not being posited in the political arena in any truly substantial way. Despite the great improvements in the Proton Exchange Membrane fuel cell area, the great investment in "alternative" vehicles by Ford and Daimler-Benz, the marketing of Japanese-made bi-fuel vehicles, the use of natural gas by United Parcel Service, and the U.S. Department of Energy's Partnership for a New Generation of Vehicles (or the "return of the Zeppelin"), the commercial realization and widespread adoption of these alternative technologies seem as far off as ever in terms of volume impact—and these alternatives have been tried many times in the past.

Rationing

When something is scarce, precious, and expensive but at the same time fundamental—especially in the short run—to the very basis of economic and domestic life, then in a time of crisis there would be some official intervention to conserve the remaining supplies so that we may better weather the crisis. We came close to this in 1974, and several countries—the United Kingdom, for instance—actually distributed "petrol" ration books to the general public. The books, however, were never used. In the United States, ration cards were prepared but not distributed. It might be argued that in most of the developed world outside of the United States, there is a form of de facto rationing by virtue of the fact that gasoline prices bear extremely heavy taxes—often in excess of 70 percent or more of the total retail purchase price (several hundred percent of the landed price of the oil). In the United States by 1995, taxes constituted $.42 per gallon of gasoline. However, the overall price of crude oil had dropped so much that the total price paid at the pump was considerably less than it was in 1981, when taxes were only $.28 per gallon. So, despite a substantial increase in taxation on gasoline, the public did not notice because it was part of an overall *declining* price at the pump during the same period as crude oil fell from a high of $1.40 per gallon in 1981 to $.41 per gallon in 1995.

A dramatically higher price for fuel results—or should result—in smaller, more fuel-efficient cars; possibly lies behind the more abundant public transport alternatives; and helps account for the fewer miles per annum traveled by the average European compared to the average American, although distances are much shorter there.[2] However, this is scarcely a form of rationing, strictly speaking, nor is it a conscious policy of energy conservation and management—rather, it is a means of raising vast amounts of money for general revenue purposes from motorists and vehicle users in a remarkably unresponsive market, putting them up there with the smokers and drinkers. It is attractive to argue, along the lines of an unintended consequence, that these huge fuel taxes are an instrument of *conservation,* and it is true that they prevent the squandering of energy resources, but the British tax, for instance, was 75 percent of retail price back in the 1950s, long before anyone considered conservation. In a chapter dealing with policy we must be clear about why a thing is done and not what fortuitous circumstances may result from it. The gasoline tax has been a golden goose for general revenue for many years. But, high prices through taxes do make public transport, for instance, a more attractive alternative and make you think twice about taking a journey.

In the United States, recent energy tax increases have been specifically for the reduction of the deficit, not for conservation purposes. Europeans, with their tradition of higher taxes, will tolerate a conservation tax (see Figure 5.3); Americans will not without a very visible real or imagined crisis. This, in part, reflects the much higher density of population in Europe, the much shorter average journey, and the presence of public transport alternatives. The alternative to the car in much of Europe is the bus or the train. In the United States, it is the plane—just compare domestic airfares in the United States and in Europe.

Furthermore, in the face of scarcity and even the possibility of rationing, we would assume that the consumer would seek options that provide for the more efficient use of the scarce resource. Instead, what we see now in this country is

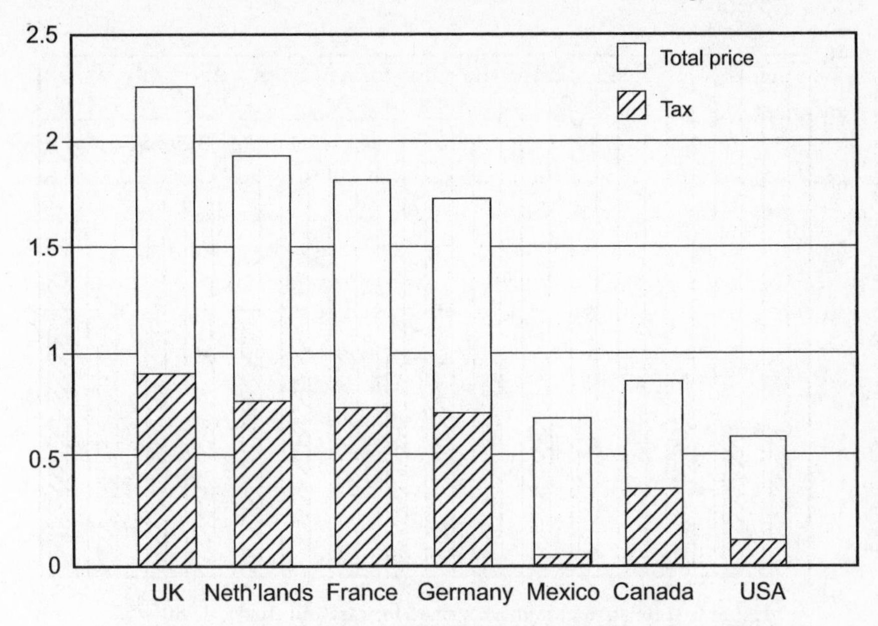

Figure 5.3. Gasoline cost per liter in approximate U.S. dollars, 1996.
Source: American Petroleum Institute.

the *increasing* size of the typical American road vehicle—sport utility vehicles (SUVs) and trucks now account for more than half of new private vehicle sales in the United States (see Figure 5.4)—the growth in the size and output of vehicle engines, and a general inapplicability of the Corporate Average Fuel Economy (CAFE) guidelines for vehicle fuel efficiency that refer to cars but not to SUVs, trucks, vans (the three categories now dominating sales), and the like. Congress enacted the CAFE standards, which specify target miles per gallon, under President Ford in 1975. The average car today is, at twenty-eight miles per gallon, twice as fuel efficient as its counterpart of 1974. However, cars are yielding to SUVs, vans, and trucks, with an average fuel efficiency of around fourteen miles per gallon, halving the target saving in gasoline and making the CAFE guidelines less significant.

Consider, for instance, the Ford Motor Company's introduction in 1999 of the latest version of the Expedition, at well over two tons the largest passenger vehicle they had put on the market in decades, or the introduction of the MOG by Mercedes, which is the largest passenger vehicle offered to the American public since the 1960s. Meanwhile, sales of subcompacts languish, and the cost per mile of travel in the United States in mid-1999 was at the lowest point it has been since 1935.

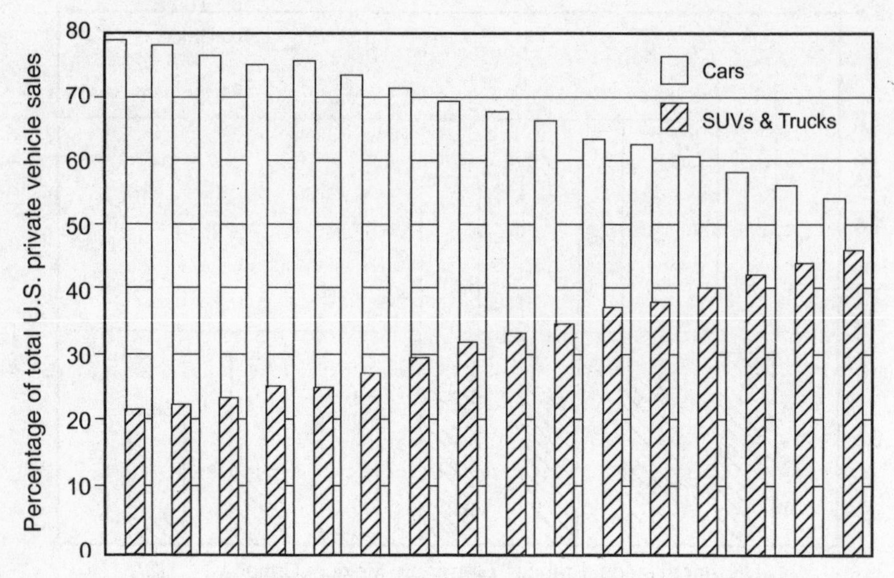

Figure 5.4. Relative U.S. market share for cars and trucks, 1980–95.
Source: American Petroleum Institute.

The Problem of Public Perception

So, the simple fact is that the public, operating within this set of "signals," perceives no crisis, whatever the scientists may tell them about the lifetime of reserves. The indicators are simply not there "on the street." We do not, for instance, have an Energy Czar as we have a Drug Czar. In the absence of a groundswell of deep public concern, we would normally not expect to see too much pressure for change, and this issue certainly did not become a central—or even significant—feature of the 2000 presidential debate. The impetus for the electric vehicle (the passage of legislation for a minimum percentage of "zero-emission" vehicles) in California, after all, was due to the *pollution* effect of the internal combustion engine, not to the fact that the engine was burning nonrenewable gasoline.

There is also a difference between the concern that led to the environmental policies, instruments, institutions, and so forth of the 1960s and 1970s, and the "energy crisis." In the environmental case, there was a very definite "them and us" context to the debate. The perception was that the birds were being killed by "industrial" pesticides, the rivers were being killed by "industrial" chemicals, the air was being choked by "industrial" emissions, and those same emissions were killing forests, acidifying lakes, and so on. What all this had in common was that the general public—or at least a substantial vote-influencing part of it —perceived itself as being in the position of a "victim" of the relentless pursuit of short-term, sinister, profit-taking capitalism that passed on pollution costs and pollution effects to the community at large. Such a polarized situation makes for a good policy "fight," so it was quite possible to construct the image of "Joe Citizen" against the combined forces of giant capitalism—whether or not this was the reality. Nothing so embodies this conceptualization as the depiction in several documentaries of Rachel Carson as the small, sickly, frail voice of honest science and the conscience of nature, being taken on by the silky-voiced emissaries of the agrochemical business, losing more and more ground every time they tried to belittle her status.

However, in the current policy context, the main culprit of energy depletion and pollution is "Joe" himself, as long as he enjoys the benefits of his car (1999 was memorable as the year in which cars finally outnumbered people in the United States), his centralized home temperature control, his travel by jet plane for recreation, his commute from the suburbs alone in his car, and his second home, third watch, twentieth pen, and so forth. In considerations of energy, we all too frequently forget the enormous amount of energy it takes to *make* things—and this is the age of things. In 1935, the average American male had three changes of clothes and certainly did not own a car. A watch and a pen were purely functional items, not fashion accessories. There can be little finger-pointing, any longer, at corporate America on the energy issue, for transportation accounts, for instance, for more than half of the oil consumption in the United States—and the fabric of our lives is woven around this contemporary lifestyle: Joe must point the finger at himself.

In other words, a serious energy policy requires an equally serious recognition of personal responsibility and a willingness for everyone to, as it were, "reform themselves." Of course, the rich consume more energy in a consumer society, but even what we might describe as "modestly comfortable" Americans not only consume staggering amounts of energy (the average American consumes approximately eighty times the amount of energy as the average Bangladeshi), but anywhere up to 65 percent of it is used at extremely low levels of thermodynamic efficiency. This might be a good point to mention the concepts of *conservation* and *waste* of energy. It is often said that the most dramatic savings in energy in this country would come from the reduction of "waste" and a conservation approach. We may make gasoline engines more *efficient,* in the sense of yielding more miles per gallon (efficiency being an input-output ratio), but the question is: is the personal car, especially a giant SUV, simply not a wise use of resources when we could move ourselves around perfectly well without using so much energy? Are we not using energy for very marginal purposes (security lights, keeping the television "live," owning so many things we could easily do without, and so forth)? This is where waste comes in. We could probably live very comfortable, meaningful lives without 30 percent of the energy we consume now.

In the absence of any sense of crisis, it is unlikely that the public at large will be ready to initiate and accept the changes to lifestyle that are required to counteract the present position of extravagant energy consumption to maintain fashions and styles of life, consumerism, and indulgence. A further insight into "extravagance" arises from a consideration of thermodynamic efficiency. For example, coal-burning power plants in the United States are less than 30 percent efficient; from oil well to moving car, we waste over 90 percent of the energy contained in the oil in the ground if we consider energy used that does not achieve the primary purpose of forward motion.

One writer has said that "three things need to be present for an item to become prominent on most agendas: *scope*—it affects a lot of people; *intensity*—its impact is sharp; and a *triggering mechanism*—a dramatic event to spark action."[3]

Part of the problem is that, having built a society and economy on cheap energy and a high order of personal mobility, it is difficult to alter the infrastructure that such phenomena engender to suit another system. If, for instance, even if we all decided to use public transport, the extremely low settlement density of modern suburbia would make an efficient (that is, customers per mile) public transport network costly and difficult to develop as well as slow and frustrating for the passengers—particularly away from the conurbations. Similarly, our scattered, individualized configurations of dwellings make it difficult to provide the sort of district-level heating and cooling that is possible in apartment buildings and high-rise residential areas.

Policy Trends

What passes for energy policy in the United States has changed over the years, and we may recognize several epochs: the age of the engineer, the age of security, and the neutral or environmental period. This relates to the way in which we have seen energy—sometimes as a means of conquering nature, sometimes as the fountain of eternal betterment of life, and sometimes as a commodity with some possibly serious "environmental consequences."

The Age of the Engineer

The earliest elements of policy were to assist in the exploitation and development and distribution of energy. This was achieved by massive public works

projects, such as hydropower dams, as well as the re
Tennessee Valley Authority, the rural electrificatior
tion of the national highway system, and airport
the heroic sounds of the conquest of nature in the
tions and of the mighty progress of modern man. T
1960s.

The Age of Security

Starting with the first Teheran oil price rise in 1969, and then through the tumult of the "oil shocks" of the 1970s, there was more of a preoccupation with where we got our energy—specifically, oil. The strategic exposure of the United States and Europe was dramatically revealed. Under the Nixon and Carter administrations, various conservation and strategic measures were introduced, including lower highway speed limits and the strategic petroleum reserves. In France and some other parts of Europe, this potential strategic energy stranglehold led to the rapid acceleration of an alternative energy source: nuclear power, which now provides, for example, over 70 percent of France's electrical energy needs.[4]

The strategic stance also led to a dramatic acceleration in the search for new sources of fossil fuels as well as to a short-lived government investment in alternative energy research and development. The search for new sources of traditional fossil fuels, particularly oil fields, became the dominant activity—edging out alternative energy—with the North Sea, Alaska, and other fields being rapidly developed, often against serious environmental opposition. For a brief moment, a sense of true crisis had touched energy, but it was not to last long. Despite regular contretemps and wars in the Middle East, our propensity to import has never been greater, and our policy—such as it is—is to have a much stronger military response to threats to our source of supply as well as to diversify the sources of that supply across—if possible—a more secure portfolio of places. For the George W. Bush administration, that means finding new sources right here in the United States.

The Neutral or Environmental Period

Once the "oil shocks" were "over," the hysteria rapidly dissipated, and by 1986 the world was facing a glut of oil and collapsed oil prices. This, plus the Reagan era of deregulation and market economics, put most of the programs to foster alternative energy on the back burner. What Reagan did not do, Chernobyl and Three Mile Island accomplished for nuclear energy in the ates. The official perception was that most energy problems are to "market imperfections," and that perception was best dealt with by more policy but by deregulation. Under George H. W. Bush—a exan—there was great emphasis placed on the development of other domestic sources of supply to counteract the vulnerability of the United States to the Middle East.

During the 1990s, the environmental aspects of energy began to be more widely debated—especially with Al Gore as vice president. Although the Chief Executive made bold with commitments to the Kyoto meeting to reduce carbon dioxide emissions, these cannot pass muster in Congress in the face of enormous industry and labor resistance, public ambivalence, and the inevitable cost burden it would impose on a resistant voting public. There have been initiatives out of the White House promoting energy efficiencies in departments and agencies, and there have been independent state initiatives—most notably in California. However, the predominant policy in 2001 seems to be to find more oil—and the closer to home the better.

Obstacles to Confronting the Need for an Energy Policy

Where is an energy policy to come from? Reduced speed limits, CAFE, and smaller, more fuel-efficient cars came out of the OPEC embargo and the 1979 rerun of that crisis. However, since then, despite the Kuwaiti reflagging of tankers, the Gulf War, and the burning of the Kuwaiti oil fields, we have seen nothing, until the first half of the year 2000, but a relentless decline in the price of fossil fuels in the United States. Interestingly, a report produced by the International Center for Technology Assessment in November 1998 suggested that, taking into account all of the subsidies and externalities, including the military cost of securing our foreign sources of supply, the true cost of gasoline per gallon to Americans is not around $1.50 to $2.00 but more like $15.14—three times the highest pump price in the developed world.

The conservation impetus that came from 1974 and 1979 has faded away as cars have progressively become vans and then SUVs (exempted from the CAFE regulations), speed limits have been raised, people are driving more miles than ever each year (see Figure 5.5) in ever less fuel-conserving vehicles, and trains have yielded to trucks and planes.

Compounding the above problem of getting the public to accept the need for a policy-induced change are the factors that actually stand in the way of a change. Listed briefly here and discussed below, these factors include price, sunk costs, tax resistance, time horizons, and the nature of the evidence.

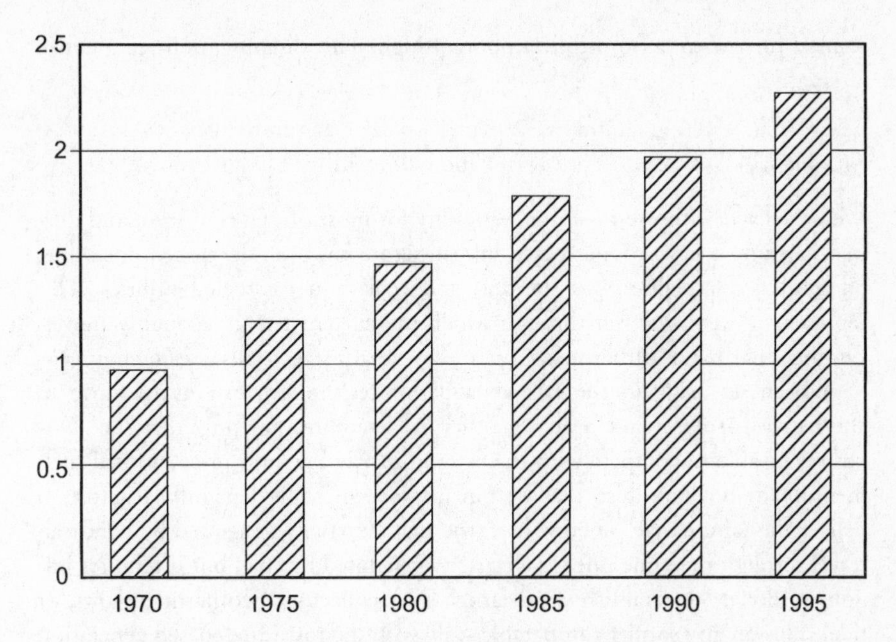

Figure 5.5. Billions of miles traveled annually by road in the United States.
Source: American Petroleum Institute.

The Low Cost of Fossil Fuels

With gasoline selling in the United States (in early summer 2001), perhaps aberrantly, at between \$1.50 and \$2.00 per gallon—having been as low as \$.73 as recently as 1998—the energy industry faces a major difficulty in moving to "alternatives" because renewable energy sources would have to compete with the extremely low base price of oil to recover their costs and turn any sort of profit. As of summer 2001, gasoline prices had been up to \$2.00 per gallon, but pressure is being exerted by consuming countries and some of the larger producers (for example, Saudi Arabia) on OPEC to *increase* oil supplies. One of the main reasons given by Saudi Arabia is that current high prices would tend to favor *alternative fuel sources,* which would be detrimental to Saudi Arabia's interests.

The bankruptcy of one of California's largest wind farms was, to a large extent, a direct result of the impossibility of competing with such extremely low oil prices in the mid-1980s. Furthermore, despite the recent rise in oil prices during 1999–2000, the prognostications in the medium term and all of the political pressures from the United States and Saudi Arabia are toward a substantial price *decline* from this temporary high. This discourages investment in alternatives.

Sunk Costs

We have lived with the fossil fuel economy for most of two centuries, and during that time a vast product-based infrastructure has been developed, including oil refineries, pipelines, gas stations, and coal- and gas-fueled utilities. This represents a vast *sunk cost* through which the current energy economy moves. On the other hand, earlier infrastructures—canals, stage trails, steam engines—have been set aside in the face of superior technology. For us to move to alternatives—solar, wind, water, geothermal—requires not only that the alternatives be able to compete with the extraordinarily low unit price of fossil fuels mentioned above but also that we put in place an equivalent infrastructure to deliver the "alternative" energy. It is true that electricity generated by wind can travel through the same lines as electricity generated by coal, but it requires billions of dollars to establish wind farms, solar collectors, geothermal plants, or tidal traps on any sort of comparable scale with the fossil fuel–based generation of energy, and that initial cost has to be compared with the established sunk costs of the present energy system, which accounts in part for the very low cost of gasoline and fuel oil. Of course, in the early days, there were huge capital costs to build this infrastructure. However, in those days, the energy business was not competing with any sort of existing huge established infrastructure, apart from canals; rather, it competed with windmills, horse-drawn wagons, and technology that had not changed much since the days of Julius Caesar. When oil took over from coal, it had obvious technical advantages, such as its easier transportation and conveyance, its greater energy content per ton, and the general advantages of a liquid over a solid for many purposes. The "alternatives" do not have any of the same manifest short-term technical superiority in this context as oil had over coal (although great hopes have been held out for the fuel cell) in the absence of high fuel prices.

Tax Resistance

If private energy corporations are unlikely to go against their established investment patterns—especially if there is no immediate public concern about an energy crisis—and if it is going to prove hard to capitalize alternative energy privately in the climate described above, in which investors see no real prospect of a handsome return, then a case would have to be made for the public sector to intervene if change is to happen without waiting for the indicators outlined above to become "critical"—when fossil fuels really do start to run out. In other words, the policy would be put in place deliberately to effect or influence a substantial, earlier systemic change from the fossil fuel economy. The argument here would be that only by state intervention to kick start the transformation would the status quo be changed more quickly, for only the state controls the mechanism to produce such a change. For instance, there could be a fossil energy tax or a carbon tax to reflect the exhaustible and damaging nature of the resources upon which we now depend and to deflect use toward more "sustainable," or "green," alternatives. The German Green movement has suggested a 300 percent increase in existing energy prices to allow for the realization that nonrenewables are an exhaustible asset. The government could, for instance, insist on all public vehicles meeting higher efficiency standards, in the way that government had earlier made provision for agencies and departments to use recycled materials, thus providing the demand that lowers the price, making the product "competitive." This case, however, is addressing the *environmental* consequences of using energy, not the source of the energy.

The problem is that the public at large, in the United States particularly, is unwilling to accept anything that looks like a tax increase, and price distortions using public money—unless they resemble agricultural subsidies—are beyond the pale at the moment. Indeed, small increases in energy taxes have been refused, and even removed, in recent years. The fact is that the United States pays less tax per gallon of gasoline than almost any developed democracy (on average, other developed democracies pay between $3 and $4 tax per gallon, as opposed to less than $.50 here), and it is not about to see that situation change. In 1996, Congress and the Clinton administration repealed the 4.3 cent surcharge to the gasoline tax introduced in 1993 as the second part of a deficit-reduction package. Teamsters and motorists alike rise in ire at any such suggestion. On the other hand, when the country at large believed there was a crisis (during the Gulf War, though supply was never seriously reduced), it paid 50 percent more per gallon almost immediately. During the conflict period, the miles driven did not drop at all, and neither did the amount of gasoline purchased. This suggests that we could easily tolerate a significant increase in energy prices—though still much less than what most of the rest of the world pays now—but we refuse to accept such a tax imposition because, simply put, we see no justification for it. It is also worth mentioning that in Europe the population density is so high, and the average journey so much less than in the United States, that it is—with much smaller cars and available public transport—possible to tolerate the extremely high fuel taxation. So, reducing energy inefficiency and "waste," fostering conservation, and encouraging clean and sustainable alternatives are all unlikely using the government-intervention-through-taxation model. Thus, it will be very hard to work through the price mechanism, presumably until energy really does become scarce and prices adjust to that. This is a case of *strategic* management versus *crisis* management—do we anticipate the "coming crisis" and prepare for it with proactive policies, or do we wait for the crisis of shortages, soaring prices, and so forth and try to deal with it then?

Time Horizons

One of the central problems of putting policy makers and "environmentalists" in the same room is that they often think according to different time scales. Time scales can, and do, radically shape our perception of reality. Consider, for instance, the following: Alternative energy groups are saying, "It is inevitable that we are going to face rapid depletion of fossil fuels, so what is the benefit in waiting around for that to happen? We need policies to anticipate that now!" Theirs is a *strategic positioning* point of view based on a rational interpretation of the scientific evidence and the laws of nature, and they could justifiably ask: who could argue with that? It seems that the answer in terms of daily reality is "most everyone." Politicians and policy makers have to operate in a world of short time horizons—usually defined by the run up to the next election—and this consideration is as influential as any other factor in shaping the policy approach to energy.

Fig. 5.6 Short-term vs. long-term view.

It is easy to be cynical, but politicians and policy makers are engaged in giving the majority of the public what it "wants," which means, what it is willing to vote for here and now. Richard Nixon understood this and also understood that, especially among the part of the population that voted Republican, there was a deep underlying concern about the destruction of the environment. To that end, he gave them the National Environment Policy Act. It is hard, even now, to think of Nixon as "Green," but he was attuned to what the voting public valued, even in the face of serious hostility from such stalwart Republican allies as big business. In contrast, the great "Republican populist" Ronald Reagan made a serious mistake when, at the beginning of his first term, he judged that deregulation, as a policy, should override the public's concern with environmental management, his view being that all of the new green regulations were constraining industry. The consequent assault on the Environmental Protection Agency (EPA) had to be stopped rapidly and turned around when he discovered that the people who voted for him were still strongly convinced of the benefits of regulating the environment. Strangely, that other great Republican populist Newt Gingrich made precisely the same mistake in his "Contract with America" and had to retreat very swiftly from his position of "rolling back" the EPA. The public sees the environment, and the maintenance of environmental standards, as an *immediate* concern that touches their lives directly—and it is hard to find any parallel concern for energy. *Indeed, the main transformations in energy policy, in the short run, are far more likely to come as offshoots of the environmental policy arena—that is, atmospheric change, pollution, and so forth—than as concerns over price or depletion.* However, the recent rejection by the George W. Bush administration to the Kyoto Protocol does very much suggest there is real reason to doubt whether the public and their political institutions are ready to change: Congress (by a demonstrated vote of 95-0) would never have ratified what Al Gore initialed, arguing that we have already invested billions in clean air since 1971 for which we should get some recognition—while other giant producers, such as India and China, remain untouched by the provisions of Kyoto.

The Nature of the Evidence

One further complication in developing any sort of serious energy policy is that the public is justifiably confused by the conflicting messages it hears on the subject of an energy crisis—real or imagined—and has become to some extent distrustful, or even cynical, with respect to the information it receives. The enormous flow of data in the information age does not necessarily clarify the policy process any more than it helps select the "best" long-distance phone carrier. This is especially the case when, as we have observed, all of the standard "crisis" signals are not there on the public's horizon. Not least among these conflicting messages is the continual upward revision of the "proven reserves" of oil and—in the general public's perception anyway—the seemingly vast reserves of coal, and their "fatigue" response to frequent, and unrealized, prognostications of the doomsayers. For instance, it seems to the layperson that a "forty-year reserve of oil" has been out there for at least the past fifty years. In a November 1994 article entitled "It Costs More to Save Energy," the *New York Times* commented:

> Frederich von Hayek . . . noted that conservation can sometimes do more harm than good. "Industrial development would have been greatly retarded if, 60 or 80 years ago the warnings of conservationists about the threatened exhaustion of the supply of coal had been heeded," he wrote in 1960. "The internal combustion engine would never have revolutionized transport if its use had been limited by known supplies of oil."[5]

The coal reserves, of course, would appear far less substantial than they sound at present if we shifted the demand for oil toward coal. When we look at the unreliability of the doomsayers, we go back to a problem that has been with us since Thomas Malthus at the end of the eighteenth century. His predictions of famine, pestilence, starvation, and the like did not come to pass in the way he imagined. Indeed, the world "supports" unimaginably more people now than it did in Malthus's day. The "X-factor" in all of this was technology. No one can argue that Malthus got it wrong. He simply could not have taken into consideration—any more than we can now—things about which human intellect, at that time, was totally ignorant: the technology of the Green revolution, transportation, refrigeration, gene splicing, and so forth. Malthus was reborn as "Limits to Growth," a document that had enormous policy exposure through the Club of Rome.[6]

On the other hand, it can be argued convincingly that the exercise in "Malthus with computers" was no more accurate than was the reverend economist himself, and most of its projections turned out to be well off the mark—perhaps because the pace of technological and social change has itself accelerated so much since Malthus rendered forecasting such a chancy business. That leaves us, in the policy arena, with two camps: Micawber and Cassandra—the former reassuring us that "something will turn up," transforming technology into something akin to a religious belief, an act of faith; the other telling us, to our total disregard, that "the end of the world is nigh" and we have only been putting off the evil day. (Cassandra, in Greek mythology, was the daughter of King Priam and Queen Hecuba of Troy. The god Apollo, who loved Cassandra, granted her the gift of prophecy, but when she refused to return his love, Apollo made the gift useless by decreeing that no one would believe her predictions. Wilkins Micawber is a character in Dickens's *David Copperfield* (1850) who is poor and somewhat idle but lives in the constant expectation of things getting better, though not necessarily as a result of his actually doing anything to change events.) All of this is encapsulated in the popular works by Julian Simon and Paul Ehrlich, in which huge volumes of "evidence" are paraded before the general public to "prove" totally irreconcilable interpretations of the effect of geometric population growth.[7] Politicians are most unlikely to take initiatives in this type of policy environment.

From the above, we can speculate that the policy arena is a difficult melange of those who say, "Let the hidden hand of price, the market, and scarcity rule the day—it has taken care of us in the past"; those who say, "Why not be rational and farsighted and anticipate the inevitable?"; and those who really do not say anything at all because life is fine the way it is.

Clouding the policy issue even more is the fact that the most likely "alternative" energy source—nuclear energy—is perceived by a substantial part of the American public (if not the French) as dangerous and undesirable, especially since Three Mile Island and Chernobyl.[8] Thus, the principal policy alternative is, seemingly, ruled out as an option in the largest energy-using economy on earth, and the other alternatives are nowhere near being ready to take over.[9] In short, there really is no strategic impetus for the preparatory long-term energy policy changes that some environmentalists would have us begin to initiate now. This, however, is despite estimates that the energy potential of high-quality wind resources, allowing for siting issues, are around three times the present world electrical energy use.

Energy, the Environment, and Policy

As mentioned above, from a policy perspective, more response is likely to come from, and be generated by, the public in terms of perceived links between energy and the environment than by an energy "crisis" in and of itself. There is nothing to suggest a fear of such a crisis at present.

On the other hand, the language of crisis is certainly being invoked by some in the scientific community. In an October 1998 issue of *Nature,* scientists from Europe and North America "predicted that global warming will soon become the environmental equivalent of the Cold War as the increasing reliance on fossil fuels releases more carbon dioxide and other heat-trapping pollutants into the atmosphere."[10] Martin Hoffert, a physicist at New York University, commented: "Developing and commercializing carbon-free power technologies by the mid-

21st century could require efforts, perhaps international, pursued with the urgency of the Manhattan Project or the Apollo Space Program."[11]

Not only is the language of the article unusually speculative for a vehicle of scientific objectivity, but the article itself is unusual inasmuch as it ventures into the area of policy, which generally is an unusual focus for scientific organs and scientists. In some respects, it seems that science is running well ahead of policy. Despite such observations—the Kyoto meeting notwithstanding—countries have *done* little to influence the causes of atmospheric change. Some scientists are proclaiming that global warming is inevitable and that no amount of effort—not even a crash program—will prevent it. Martin Parry of University College, London, stated: "We *will* experience a substantial amount of further climatic change even if we make *huge* cuts in emissions."[12]

Undeniably, there is serious public contention among the scientific community over the extent of the effects of global atmospheric change and the degree of human responsibility for causing the change. These types of statements from the scientific community, if read by the public at large or by the government, tend to reinforce the position that "there's nothing we can do about it anyway"—neatly complementing the "they can't seem to agree on whether anything is happening anyway" stance. Neither of these is likely to lead to policy and funding for a "Manhattan"-scale or "Apollo"-scale project as mentioned above. Furthermore, there are *huge* forces representing the status quo ranged against any systemic major adjustments to the current energy equation. Their views are clear—the voting public really seems not to have any strong cohesive stance on sustainable energy.

That is not to say that public perception may not change mightily and fast. It is a feature of global atmospheric change that the impact is felt not so much through moving averages as through "extreme events."[13] What this means is that, instead of the slow and steady change that the term *global warming* implies, the huge amounts of energy being retained in the atmosphere will work themselves out in terms of violent short-term events, such as typhoons, droughts, floods, storm surges, and the like. They have a strong visual impact, and, in the case of tornadoes and hurricanes in the United States, they have a dramatic overnight impact on the insurance industry. The billions of dollars each event costs the industry will show up in premiums, cancellations, conditions of "non-insurability," and so forth. The insurance industry may well provide better barometers of crisis—through risk assessment—than does the public at large, as long as it is not sheltered from reality by government intervention. In its publication *Vital Signs,* the Worldwatch Institute stated, in solid layperson's terms:

> Last year's [1998] record average temperatures literally went off the top of the chart we have been using for years. . . . The

> unprecedented warmth caused more evaporation and rainfall than usual, and provoked more destructive storms. Weather-related damage worldwide totaled $92 *billion,* 53 per cent more than the previous record—the $60 billion recorded in 1996.[14]

The real difficulty from the policy point of view is not just making the connection between these extreme events and our energy-use patterns but coming to the conclusion that these individual events are not, simply put, acts of God or natural disasters, however disastrous they may be. The policy response to an act of God is based on a sense of the inevitable, immutable, and unavoidable; to a natural disaster, the policy reaction is usually better response preparedness and postdisaster *relief.* None of this tackles causation in energy-use terms—not surprising, considering that there is such contention among the scientific community.

So, we may look to environmentalism as a surrogate mother for the child called "energy policy." This approach addresses not the sources of energy but the consequences of using certain types of energy. Yet, the adverse effects of most current forms of energy conversion encourage a strong environmental emphasis on conservation and zero-emission alternatives that might move us toward renewable energy sources. Furthermore, environmentalists encourage us to look at the "real" costs of using fossil fuels (incorporating environmental externalities); this, too, might encourage new pricing policies.

Policy for Whom?

In this discussion, although I have concentrated mostly on the U.S. policy environment, I do not intend to review this country's energy policies in detail. In essence, the central problem emerging now is that it is perhaps illusory to draw a distinction between "national" and "international" policies regarding energy. The sources of supply are hugely international, as indeed are the global consequences of the use of energy—however regional may be its use. So, on the one hand, we need global policies to handle global problems, such as climate change and other atmospheric phenomena, but we have no global government or—strictly speaking—international law, in the sense that we understand law within the confines of a sovereign state. States choose to be bound to international law and can walk away from it if it displeases them, as the United States did over the mining of Nicaraguan ports. We have pronouncements, treaties, and protocols—many seemingly unenforceable and all lacking serious *real* sanctions in the face of individual state failure to meet obligations and stated responsibilities. On the other hand, sovereign nations have to make decisions that ultimately will affect many—

or all—parts of the world. This is particularly true for the industrial giants, who, while accounting for only 25 percent of the world's population, consume more than 70 percent of the world's energy, with a corresponding pollution effect.

Oil shocks apart, energy policy traditionally has been concerned with *developing* energy by building grids, expanding fuel recovery, and so forth. The term *energy policy* came to mean (during the Nixon and Carter administrations) "reining in" the cowboy activities of the past, and this was seen as a threat to competitiveness, growth, and "progress"—just as the environmental movement was perceived in this way in the 1960s by steel and agrochemicals, among others, to have the same negative, anti-industry caste. This perception is illustrated by the following quote from a speech by C. Papoutsis to the 1995 Brussels Conference on European Union Energy Policy.

> Competitiveness is an essential element of the economic future of the [European Union]. We all recognize the importance of competitiveness for maintaining jobs, welfare and quality of life. Energy policy cannot be considered outside this context, as an isolated case. The energy sector has to participate in the [Union's] efforts towards improved competitiveness. Industry is rightly concerned with this question.[15]

In fact, the environmental policies enacted after 1969 created a vast new industry of pollution control, in which the United States is a world leader.[16] That experience does not seem to have influenced the energy debate, and—apart from nuclear alarm—the public is not set to take on the status quo. Under the present Bush administration, energy policy seems to have been captured by the notion of expanding supplies—particularly domestic supplies—of conventional fossil fuels with a heavy emphasis on oil. There was a deep reduction in alternative energy funding and research in the 2001 federal budget proposals.

Conclusion

At present, there is no pressure for a radical and different comprehensive energy policy based on assumptions about the viability of our current mode of energy use, although there are many solid and, to some people, convincing arguments why there should be one. These arguments include the following:

- *Economic productivity*—improving economic efficiency of the energy supply and end-use systems to enhance overall performance of the U.S. economy
- *Oil supply vulnerability*—reducing the vulnerability of the U.S. economy to strategic disruptions in so basic a commodity

- *Systems reliability*—ensuring energy systems reliability, flexibility, emergency response capability, and risk management
- *Managing environmental change*—reducing pollution
- *Controlling extreme events*—reducing greenhouse gas emissions
- *Continuity*—enhancing global sustainability.

The U.S. public perceives energy as cheap and readily available—and when it isn't, that this is the result of political ineptitude, bad management, political chicanery by foreign governments, or some other very human failing. There are, as we have said, two dimensions to this problem in the public mind: is there plenty of energy available for my needs, mainly today, and is there really some dire consequence to my huge consumption of energy in terms of the environment (again in the near future)? On both counts, the information being signaled to the public at large is "business as usual." These are the realities, in terms of policy, with which we have to deal. No amount of heart searching over the Bush administration's rejection of the Kyoto Protocol means anything as long as the vast majority of Congress would not give ratification the time of day—any more than they would have when it was initialed by the previous administration. Policy is pragmatic, short term, emotional, and sometimes dissonant in terms of "what I want" versus "what I suspect is good for everyone." This is where those concerned about the viability of our existing pattern of energy use and its consequences have to start.

Key Ideas in Chapter 5

- Policy in a democratic society reflects the prevailing values and aspirations of the voting public.

- There is a need for widespread public concern embodying a perceived need for change before anything can be accomplished in the policy arena.

- A sense of crisis is frequently a requirement for mobilizing public concern and political action.

- In the absence of clear signals for an energy crisis, energy policy is more likely to be generated by the public perception of the links between energy and the environment than by an energy crisis per se.

Notes

1. David Howard Davis, *Energy Politics,* 4th ed. (New York: St. Martin's, 1993), 290.
2. The United Kingdom announced in July 2000 that it plans to spend upward of $240 *billion* on transport over the next ten years—half of that goes to improving conditions for the private and commercial motorist and the rest to public transport.
3. Richard Ryan of the University of San Diego, Calexico, made this point in an address to the International Section of the National Association of Schools of Public Affairs and Administration in Miami, 1989.
4. It should be mentioned, however, that in summer 2000, the German government proposed a thirty-year phaseout of nuclear power.
5. Herbert Inhaber and Harry Saunders, "It Costs More to Save Energy," *New York Times,* November 27, 1994.
6. Donella H. Meadows, Dennis L. Meadows, Jorgan Randers, and William W. Behrens III, *Limits to Growth* (New York: Potomac Associates, 1972). The Club of Rome was the NGO (Non Governmental Organization) that sponsored the study.
7. Julian L. Simon and Herman Kahn, *The Resourceful Earth: A Response to Global 2000* (Oxford: Basil Blackwell, 1948). Paul R. Ehrlich, *The Population Bomb* (Cutchogue, New York: Buccaneer Books (reprint), 1968).
8. Though, strictly speaking, it too is dependent on nonrenewable fuel supplies as long as we are using a fission technology.
9. In mid-1999, House and Senate appropriation committees cut Department of Energy programs for renewable energy and energy efficiency programs in the fiscal year 2000 budget.
10. M. Hoffert et. al., "Energy Implications of Future Stabilisation of Atmospheric CO_2Content," *Nature* 395:6705 (1998).
11. Quoted by the Associated Press on October 28, 1998.
12. Quoted by the Associated Press, October 28, 1998. The italics are mine.
13. Global atmospheric change is a much bigger issue than global warming, which is how the issue is conventionally described.
14. Lester R. Brown, Michael Renner, and Christopher Flavin, *Vital Signs 1998: The Environmental Trends that are Shaping Our Future,* Worldwatch Institute (New York, Loudon: W. W. Norton & Co., 1998). The emphasis is mine. This is a huge rise in damage costs, and it is significant that the previous highest year was only two years previous.
15. From a speech by C. Papoutsis, member of the European Commission, to the Conference on European Union Energy Policy, Brussels, June 22, 1995.
16. It is estimated that somewhere around 30 percent of the cost of manufacturing capital these days constitutes some form of environmental control and management.

Chapter 6

Energy and Sustainable Economic Growth

LLOYD ORR

In all modern societies, human preferences and values are expressed through private institutions, such as the marketplace, as well as in the political sphere. Decisions, plans, and policies implemented in either the private or the public sphere are frequently intergenerational in their effects. Examples include individual and charitable bequests, saving for a child's education, social security, and environmental policies. However, Chapter 5 makes it abundantly clear that time horizons in both spheres are typically very short relative to the long-term perspective required by the "rules of the game." Understanding the physical, human, and political dimensions of the energy-environment problem, how can we reconcile continuing economic growth with the laws of nature? How can we implement an energy policy framework that will move us toward sustainable growth?

In the following chapter, Lloyd Orr addresses these questions. Resolution requires an understanding of the nature of wealth, markets, market failure, and the interaction of policies with complex national and international economic systems. Policies that ignore complex institutions and human interactions are likely to become costly and ineffective. They waste the very resources that it is our intent to conserve and effectively use to meet human needs and wants of the present and future generations.

—Editors' note

Can the economic and social opportunities presently enjoyed by developed nations be passed on to future generations? Can they be acquired by developing nations? The optimist considers the basis of our prosperity to be the capital, technology, and social institutions bequeathed to us by previous generations, the ongoing progress in knowledge and technology, and the growing prosperity of those less developed countries adopting political and economic institutions that support increasing per capita incomes. While not ignoring resource depletion, the optimist concludes that the process of technological adaptation to scarcities can continue for future generations and spread to nations that remain poor. The pessimist stresses the dependence of this process on the increasing consumption of exhaustible energy and mineral resources and on degradation of the environment. These activities will deprive future generations and the least developed nations of the use of those same resources for their own sustenance.

The views of the pessimists were symbolized by the controversial simulation models of the early 1970s published in *The Limits to Growth.*[1] The intermediate projections in this report have not been validated. In contrast to projections of at least the beginnings of serious worldwide degeneration by the beginning of the twenty-first century, it is generally true that real prices of energy and mineral resources are lower, energy efficiency and standards of living are higher in most of the world and, in the developed countries, there is less pollution damage than was the case thirty years ago. This does not justify a claim of victory by the optimists. As emphasized by the 1970s critics of the model, it does suggest that opportunities and complexities are greater, social and economic institutions are more adaptable, and the relevant issues are more complex than envisioned by the limits to growth scenarios. However, if the projections of these models are labeled as a failure, an important factor may be that the continuous, but stumbling, social response to the environment and resource problems, as perceived in the 1960s and 1970s, has been much more effective than anyone could have predicted. Social response was the clarion call of the authors of *The Limits to Growth.*

Sustainability, with sustainable energy at its absolute core, is the preservation for future generations of a set of economic and social opportunities that are at least as rich and diverse as our own. The definition is necessarily vague since we know relatively little about the specific institutions and resource structures that could continuously achieve this goal of richness and diversity over time. In fact, sustainability is not a specific *goal* so much as it is a *process* of continuous change and adaptation.

The discipline of economics concentrates on the central question of resource allocation among alternative ends when faced with constraints. Given resource limitations *(scarcity)* how can we effectively choose to use those resources within the present generation and between present and future generations? This question is at the heart of the sustainable energy problem. It is a very important and broad application of economic principles employing the most fundamental scarcity constraints. The constraints are those imposed by the laws of thermodynamics, characterized as the *throughput* of energy and material in an economic system with only the sun as an external source of energy—the one-way flow of energy described in Chapter 1.[2]

The term *production* suggests creation, and *consumption* suggests destruction (or the "using up") of material goods. These common interpretations are especially misleading in the analysis of energy sustainability. The throughput of energy and its embodiment in material informs us that both production and consumption are transformation processes in which services that maintain and enhance human life are provided. This thermodynamic transformation process is from low-entropy energy and material resources to high-entropy *residuals,* the leftovers from the production-consumption process. *Pollution* occurs if these residuals significantly damage the environment and directly or indirectly affect the quality of life. There is no one-to-one relation between residuals and damage. It is therefore the damage that is the pollution, not the residuals themselves.

This broad view of the economic problem requires broader definitions of some common economic concepts than is customary. *Income* is *not* one's biweekly paycheck or simply the result of market activity. It is the *flow of services* (such as nutrition, shelter, environmental, transportation, education, and other investments in productive assets) resulting from the energy transformation process. Its major components are *consumption* and *investment,* which are the provision of current services and the enhancement of wealth. The value of these services determines the value of the biological and physical artifacts that produce them, not the reverse. Thus, the value of our workers and all other resources (human, natural, environmental, and physical capital) is determined by our valuation of the flow of services they produce. In the broadest sense, these resources represent our personal, national, and international *stock of wealth.* A bicycle is part of our stock of wealth, which is measured at some moment of time. It provides income and consumption as a flow of services, in the form of transportation, over its productive life. Its initial production and purchase are an investment. Ecosystems are part of our stock of wealth in that they provide flows of essential services. Since most of these services are not traded in markets, establishing value is much more difficult than in the case of bicycles.

In an editorial meeting Randall Baker noted that an economic system, in its personal, private, and public spheres, is importantly a "mechanism for expressing and codifying what we 'value' at any given time." If you bear this quote in mind, you will share Baker's insight that this is a key to integrating the various perspectives represented in this book. We will elaborate and use the above basic concepts throughout this chapter in an effort to translate the physical and biological science into the economics of energy sustainability.

Economic Systems and Economic Planning

To formulate an effective energy policy, it is necessary to develop a basic understanding of how modern economic systems work. The misunderstanding or misuse of some basic economic concepts is responsible for numerous policy failures in terms of disappointing and frequently perverse results relative to the policy goals. A useful example of how a modern economic system works is an analogy with rush hour traffic in a metropolitan area.

Consider a structure of highways and roads with millions of people who have goals of being someplace other than where they are. A set of imperfectly enforced rules on speed and right-of-way exists as well as civil and criminal penalties for egregious behavior. The citizens are turned loose to pursue their goals. One would expect chaos, and indeed, if you are caught in this accelerating-slowing, lane-changing, starting-stopping milieu, you might very well perceive chaos. But if you are in a helicopter high overhead, you will observe the amazing implementation of an incredibly complex "plan"—an astounding feat of social cooperation that we notice only to the extent that it fails. Of course, the plan is not perfect; accidents, delays, and traffic jams occur. But, on the whole, the system functions well. Even the helicopter may be part of an effort to provide information to eager listeners that minimizes the disruptions caused by imperfections.[3]

Now, imagine trying to plan explicitly the behavior of these same citizens for reaching their goals. In this exercise, one begins to understand the overwhelming nature of the task in specifying time of departure, time en route, and roads and lanes to be used for each individual at each moment during the trip. The outcome of inevitably crude attempts to implement this plan—if enforceable—is going to be piles of mangled metal and associated human trauma. In this analogy are the beginnings of an understanding as to why a complex modern economy needs decentralized decision making under rules that provide individuals with the incentives for socially productive outcomes. Alternatively, it is also a crude but pointed lesson as to why centrally planned economies have fostered authoritarian governments and ultimately failed.[4]

Markets

How do we induce strangers to undertake the complex and interdependent actions that result in the multitude of tangible and intangible goods that we consume? This is a process of decision making in a situation that requires continuous mutual adjustments by individuals who hold legally enforceable property rights and who make decisions based on expected benefits and costs to themselves. This is not simple selfishness, which is the pejorative term used by those who wish to disparage markets. The economics of a Mother Teresa and an Ebenezer Scrooge, the head of a charitable organization and a CEO, a dedicated public servant and a hidebound bureaucrat, are all the same. They all have goals, and they all make decisions, under a set of rules, that are based on expected benefits and costs to themselves even when they represent commercial or noncommercial institutions. Regardless of personal motivation, they all face the problem of choosing how to use limited resources. Markets facilitate the achievement of diverse goals by individuals and institutions populating a complex society.

Consider the market for a simple commodity, such as bread, sold in New York City. It takes only reflection to realize that there are hundreds of varieties of bread and that the quantities of each variety sold change by day, week, season, and neighborhood. Despite the fact that no one knows what any of these quantities actually are, New Yorkers typically find the bread they want at the time they want it. This is the outcome of the efforts of thousands who are experts in some small niche of a process in which production, transportation, and distribution are coordinated by the interactions among these self-interested strangers. With countless mutually advantageous trades, they consummate well-timed deliveries with millions of final consumers. Analogously to rush hour traffic—even for such a simple commodity—no explicit planning process could perform this function nearly as well or at such a low "planning" cost.

In the commercial sector, the endeavors of individuals to make the best possible use of their resources is guided by a system of prices. Prices are governed by the laws of supply and demand. *Supply* represents the availability, or scarcity, and relative cost of providing a resource or final product. *Demand* represents the needs, preferences, and willingness to pay of those who wish to acquire a resource or final product. The interplay of markets and the resulting costs of the complex decentralized decision making involved in obtaining and combining the global resources required to produce and distribute a loaf of bread are (imperfectly) communicated in a single number—for example, $1.59. Incredibly complex and valuable information is communicated through this single number. The price of a commodity typically contains the information required by the individual and society to efficiently allocate limited resources. Thus, the effectiveness of the *market system* is seen as a decentralized planning process. The fluid coordination in the process of achieving complex goals is based on the *self-interest* of producers and consumers and on an effective low-cost *information flow.*

Major exceptions to this rosy scenario are, of course, the subject of this volume. Environmental pollution and allocating exhaustible energy and material resources over long periods of time are examples of actual or potential *market failure.* However, to deal effectively with market failures, we first need to have a thorough appreciation of how markets work.

Markets are the coordinated outcomes of a multitude of individuals making choices that conserve the limited resources they control. This is the result of an incentive structure that leads individuals to pursue effectively the goals that interest them. Their ability to achieve these goals is enhanced by an effective, continuous, and low-cost flow of information. Price signals ultimately reflect the resource costs of their decisions. Because of their limited command of resources, these signals lead individuals to make productive substitutions of less costly (less scarce) resources for more costly resources. One of the greatest advantages of markets, in comparison with other known methods and institutions for rationing scarce resources, is their ability to respond to the ever-changing relative scarcities of marketed resources as communicated by price signals. Markets work well not because of selfish behavior in any narrow sense of the term but because the constraints on selfish behavior are so effective and productive relative to the constraints imposed by other mechanisms for making economic decisions.

In formulating effective environmental and sustainable energy policy, an understanding of the interactions of policies with the interconnected system of markets is essential for long-term success. Ignoring its complex incentive structure results in damage to the economy. More importantly, it leads to policy attenuation and failure, labeled as the "law of unintended consequences." Making use of this same incentive structure can strengthen both the policy outcomes and the overall economic system. Bruce Johnson emphasizes this point in commenting on our willingness to seek solutions for tough social and economic problems:

> Traditionally . . . public spirited reactions to the environment, inflation, unemployment, energy, and all the other issues of the day are *result* rather than *process* oriented. If we think too many trees are being cut, we simply legislate that fewer trees be cut. If we are disturbed by inflation, we propose the adoption of price controls. If we are concerned by unemployment rates, we propose that government create additional jobs. In each case, we focus directly on the desired result rather than on the processes, mechanisms, and institutions that might reasonably be expected to lead to a preferred condition.
>
> Enter the economist as the curmudgeon with the message that good intentions are not enough to achieve desirable ends. It is the process that counts. Processes are not only important, they are critical to producing desired results. Since the economic and

> political landscape is littered with the wreckage of well-intentioned but disappointing programs, the thoughtful activist cannot ignore the economists' warnings. . . . Enough evidence is now available to suggest that the resulting frustration cannot be eliminated by means of "better" programs with "better" people running them. Instead, it is obvious that more attention should be devoted to the institutions and processes that led to the original, undesirable outcome as well as to the processes set in motion when we adopted programs to solve the problems.
>
> The economist's intellectual perspective on these matters is so similar to the environmentalist's that one wonders why a partnership between the two was not formed in the natural course of events. Environmentalists believe that the ecological system is a collection of interconnected and interdependent parts. An exogenous change in one sector or subsystem will lead to a chain reaction of effects elsewhere. Economists hold the identical view with respect to the economic system. . . . Everything depends on everything else. Moreover, it is axiomatic in economics that individuals will respond in different ways. Perhaps more to the point, prices and incentives *do* make a difference to the outcome. The lesson for [policy] is simply that the programs used, however well intentioned, will fail if the incentives buried in the institutional structure are ignored or poorly designed.[5]

Externalities: Market Failure and Third-Party Effects

For markets to perform their "economic planning" function effectively, the incentive-driven, diversified decision-making process that generates market activity must meet several conditions. One of the most important conditions is that individual decision makers must be *residual claimants*—that is, they receive *all* of the benefits and bear *all* of the costs of their choices. Although the approximation of this condition is one of the greatest advantages of markets relative to alternative economic planning institutions, it is seldom completely met in practice. Markets, like all other social institutions, are imperfect in many dimensions. When departures from the standard are substantial, there can be serious social and economic consequences, which are a form of market failure. Environmental degradation—a *negative externality*—is one of the prime examples of this market failure. With these rather strange new terms, we now move to a broader and more appropriate perspective on economics than the concentration on markets and the commercial sector with which it is popularly identified.

Figure 6.1. Externalities.

When we drive automobiles, we collectively impose costs on other people in the form of air pollution. When a business firm or a municipal sewage plant pollutes a body of water in which we would like to swim or fish, we are harmed. In both cases, costs are imposed on *third parties* who were not privy to the decisions that harmed them. As individuals, we are often powerless to prevent the costs imposed on us by mutually advantageous agreements between the driver and the gas station owner, the firm and its customers, or the municipality and its citizens. We lack enforceable property rights in the affected environmental resources. Since we *all* own these resources, in effect *nobody* owns them, and so we are all free to use them for our own purposes—often to the detriment of others.

These are examples of the *common property resource* problem. A classic case is a fishing ground that all are free to use. If the consuming population is small, relative to the size of this renewable resource, there may be no problem. But, if the population is large, the fishery may be exhausted. We may destroy what could be a continuous source of food supply under good management. It should be clear that the destruction of the fishery is not the intent of any individual fisher. In the absence of an enforceable private or social property right, and in the presence of large numbers of users, what inevitably follows is the *rule of capture*. This rule means that fishing will expand as long as labor and capital costs are covered by the catch. Any attempt to conserve that is made by individuals or small fishing groups will simply result in larger takings by others. No one has an effective incentive to conserve. A private party with an enforceable property right in the fishery would preserve its permanent value by charging rents for fishing. This would signal the value of the fish above and beyond the labor and capital cost of catching them. The fishery would not be exhausted. A public body could achieve this by following the same asset-preserving criterion in the form of an effective fishery policy, such as costly and limited fishing licenses. Effective property rights and *stewardship* replace the common property resource status of the fishery. Whether public or private, the core feature of property rights must be established. This is the enforceable *right to exclude* others from the use of a resource, except under conditions that are acceptable to the "owner."

Most environmental problems are similar to the fishery problem in that an absence of enforceable property rights leads to the waste of essential wealth, such as the atmosphere. The common property resource problem represents a failure of the economic system to express and codify what we value.

The basic insight is that *property rights lead to conservation.* To grasp this, compare the following:

- the home or car owner with the renter
- the destruction of nineteenth-century bison with domestic cattle
- the individual who would never stamp out a cigarette in his or her own driveway but who would dump an entire ashtray on a public road.

Unlimited examples of this type of conservation follow from our status as strong residual claimants with respect to things that we "own." Even pollution problems are sometimes resolved with private ownership. An example is fishing clubs in Great Britain. These clubs are able to restrict upstream sources of pollution from stretches of fishing streams in which they have historic, private, enforceable property rights. There are problems in the exercise of stewardship in both the private and public sector. Which form is preferable in a given circumstance depends on a diverse list of criteria, including effective stewardship, efficiency, history, and attitudes. However, private versus public ownership is not the core issue. Conservation comes from exclusion except under conditions that are acceptable to the owner. Effective property rights, whether public or private, are a primary key to resource conservation.

Environmental resource problems (related to their common property status) are at the dissipation, waste, "high-entropy" end of the production-consumption process we have labeled as throughput. Most exhaustible energy and material resources, which are at the concentrated natural resource, "low-entropy" end of the process, are owned by private or public bodies that are interested in preserving them for their most highly valued use. This process operates through markets and does in fact consider the needs of the future. At a later point, we will explain the mechanism through which future generations "bid" against the present generation for the use of exhaustible resources. We will also note the broad consensus among natural resource economists that these bids are consistently undervalued by the present generation. Markets, like most social institutions, tend to be shortsighted relative to the needs of the future. This creates a problem with respect to exhaustible mineral resources that is analogous to the common property resource problem described above: the present generation will consume "too much" and leave "too little" for the future.

Energy, Resources, the Environment, and Economic Growth

A core feature of John Sheffield's analysis in Chapter 2 is that the concept of sustainable energy flows involves increasing standards of living for (at least) the less developed regions of the world. Clearly, the issue of economic growth cannot be avoided. Some writers have avoided the term with such substitutions as "development," perhaps because of the negative connotations that are frequently attached to economic growth. This obscures more than it reveals since none of the basic issues of growth and sustainability is altered by changing the name. As we shall see, the issue is not between "growth" and "development" but between "physical" growth and growth in "value" or "human well-being." The latter has always been the economist's definition and is inclusive of the former. Our approach here is to describe the nature of economic growth, its relation to energy and resource use, and the forces that shape its direction. The purpose is to explore the wide range of both problems and promise that economic growth holds for energy and material sustainability.

Income and the Myth of Material Wealth

At the beginning of this chapter, we provided broad definitions of income and wealth and noted that wealth depends on the value of a flow of services (income) that our stock of wealth is capable of producing. A corollary is that *wealth does not primarily depend on its material embodiment.* This fact will be of overwhelming importance in approaching issues of sustainable energy and economic growth in both developed and developing economies. We begin with an excerpt from an economics text:

> In emphasizing the essential role of valuation in any measure of efficiency, we are also rejecting the common belief that economics has to do peculiarly with the "material": with material wealth,

material well-being, or material pursuits. It just isn't so, and the word *material* actually makes no sense when attached to such words as *wealth* or *well being.*

Of what does wealth consist? What constitutes your own wealth? Many people have drifted into the habit of supposing that an economic system produces "material wealth," like cars, houses, basketballs, breakfast cereals, and ball-point pens. But none of these things is wealth unless it is available to someone who values it. Additional water is additional wealth to a farmer who wants to irrigate; it is not wealth to a farmer caught in a Mississippi River flood. A food freezer may be wealth to an Alabaman but not to an Eskimo. The crate in which the freezer was delivered is trash to an adult but a treasure to a small child who sees it as a playhouse.

Economic growth consists not in increasing the production of *things* but in the production of *wealth. And wealth is whatever people value* [italics added]. Material things can contribute to wealth, obviously, and in some sense are essential to the production of wealth. (Even such "nonmaterial" goods as love and peace of mind do, after all, have some material embodiment.) But there

Figure 6.2. Wealth is whatever people value.

> is no necessary relation between the growth in wealth and an increase in the volume or weight or quantity of material objects. The indefensible identification of wealth with material objects must be rejected at root. It makes no sense. And it blocks understanding of many aspects of economic life. Trade (or exchange) is the best example.[6]

The author goes on to note the suspicion with which traders and "middlemen" are regarded and relates it both to the observation that the trader does not add material content to the goods being traded and to the confused association of wealth with material things. But we know that traders as wholesalers, retailers, and even speculators add value (wealth); otherwise, no one would pay them to perform their services. However, other examples will serve better to illustrate the importance of exposing the "myth of material wealth" for our focus on energy sustainability. With these examples, we will begin to see why the *Limits to Growth* model projections of the 1970s have not adequately captured the implications of economic growth (at least up until now); why the prospects for sustainability may be less bleak than many suppose; and why technology is so important in addressing the constraints imposed by the first and second laws of thermodynamics.

My "father's Oldsmobile" was a hot and well-regarded car of the early 1950s. (The self-deprecating advertisements "*not* your father's Oldsmobile" referred to a later generation of cars.) It weighed about 3,300 pounds and got sixteen miles per gallon on the road and twelve miles per gallon around town. The tires and exhaust system wore out every twenty to twenty-five thousand miles, and the battery every two years. At one hundred thousand miles, it was a rusting, bedraggled oil burner. My present car weighs about 2,500 pounds and is decisively better in virtually every dimension. It accelerates faster, handles and brakes *much* better, and has nearly the same carrying capacity. It is more crashworthy, which saves lives and medical resources. It gets more than twice the gas mileage of my father's car. Its energy efficiency alone cuts pollution by more than half. Other design features cut the damage from its energy conversion much more. Its tires last sixty to seventy-five thousand miles, and at one hundred thousand miles, it isn't even breathing hard. It will probably go in excess of two hundred thousand miles. Unlike my father's car, most of it will be recycled when its useful life ends. It clearly represents much greater personal and social wealth than my father's car. However, the *material use* per mile of service, both in ownership and in operation, is markedly less than half of the early 1950s car. This means *energy use,* since all of our material goods are just embodied forms of energy.

Think of the energy conversions—from rubber tree seed and iron ore to tire store—involved in producing and distributing a steel-belted radial tire.

The communications revolution represents a series of striking resource-saving innovations. Satellites and fiber optics are substituted for copper wire. E-mail and long-distance calling are obviously markedly superior to first-class postal services for many, if not most, purposes. Teleconferencing substitutes for business travel, saving valuable work time and material resources. Wealth is substantially enhanced. The energy and material resources required to accomplish a variety of tasks are a fraction of what they were even twenty years ago.

Other examples are numerous. My current gas furnace is 90 percent efficient. The previous one was 55 percent efficient. In the 1960s, the best thermal power plants were about 35 percent efficient. The efficiency of the best power plants is now 55 percent. The real cost of triple-pane sealed glass is significantly less than that of the double-pane glass of an earlier era. Vegetarians inform us that a good diet can be produced at a marked energy saving if it does not contain meat.

Many of these material- and energy-saving innovations and ideas owe their existence, or at least their timing, to the OPEC oil cartel manipulations of the 1970s. This episode represents an *artificial* energy scarcity induced by restrictive monopoly power. It was an extremely disruptive and damaging economic event in its sudden tripling of energy prices. Nevertheless, technological responses, such as those noted above, and substitutions, such as sweaters and insulation, were important lessons in the effectiveness of prices in fostering the conservation that will be the basis of any viable sustainable energy policy. However, in the case of OPEC, short-run and long-run energy conservation responses to high energy prices (plus the incentive to discover new energy resources) exacerbated the tensions one finds within any cartel arrangement. The effective monopoly power of the cartel collapsed within ten years. It is generally agreed that it would have collapsed sooner without the intervention of the Iran-Iraq war. Unfortunately (from the sustainable energy perspective), real prices of fossil fuels in the 1980s and 1990s were lower than they were in 1972—the year before the cartel's restrictive actions.

Another implication of this episode, noted by Robert Solow, is that "the monopolist is the environmentalist's friend, though both would be surprised to know it."[7] This is because monopolists reduce resource use by restricting production, enabling them to charge artificially high prices. We will not be proposing monopoly as a solution to the sustainable energy problem, but it is an interesting response to those who believe that monopoly power and corporate greed are the central causes of natural resource and environmental problems.

Figure 6.3. The monopolist is the environmentalist's friend, though both would be surprised to know it.

The Commercial Firm as an Intermediary

"Corporate greed" or, worse, "global corporate greed" is the favorite target of many activists who wish to solve resource and environmental problems with their particular brand of public policy. It carries the implication that someone else is responsible for these problems and that someone else will have to bear the cost of correcting them. Quite apart from noncorporate sources of massive amounts of resource use, pollution, and "greed," this politically convenient position badly misses the mark.

The business firm is an *intermediary* in the sense that it gathers resources from the environment and combines them to produce a product or service that it can sell at a price that will at least cover the cost of the process. It operates in a Darwinian environment, in which success means survival and perhaps prosperity, and failure means elimination. Consumers, and society at large, derive substantial benefits from this competition in the form of business firms constantly striving to minimize the amount of resources used for a particular output. In fact, the historically high and growing per capita incomes in the developed and developing world are dependent on this process.

However, the signals that society provides to the commercial sector are inconsistent with some of our social purposes. We signal that labor and physical capital are costly since they are traded in markets. But, as a result of the common property resource problem, we have historically signaled that the environment is free and that natural resources are relatively cheap. We should not be surprised that business firms in this milieu substitute environmental and other natural resources for labor and capital on a large scale. The same is true of virtually all other social institutions. That there is "greed" or, more neutrally, "self-interest" and undesirable behavior involved in this process is undeniable, but correcting the resulting problems requires a deeper understanding.

The primary result of this process is not "corporate profits at the expense of our environment." Over a period of time, and in the context of firms as competitive intermediaries, the outcome is that we buy our goods and services too cheaply but suffer the consequences of a degraded environment and a potential threat to future generations. We get the benefits, and we, along with our heirs, also pay the costs. A corollary is that all of us as producers, consumers, and citizens would reap the benefits and would also have to pay the costs of more conservation minded policies. The costs will not be paid by "greedy corporations." The importance of this insight is in the formulation of good policy. If we can escape the illusion that someone else is going to pay for a sustainable energy policy, we will get less disruptive, less costly, and more effective policies.

Maintaining the "global monopoly capital" and "corporate greed" perspectives as the causes of resource depletion and environmental degradation betray a narrowness of vision. They miss the point that the commercial sector, like most other social institutions, is simply responding rationally to the common property resource problem of unpriced and underpriced environmental and natural resources. Change is costly and, like other social institutions, the commercial sector will resist it. It is usually represented at the national level by its most powerful groups. The interactions of opposing interest groups create the shallow political rhetoric that we so often experience. We will all do better if we understand the fundamental problems that lie beneath the rhetoric.

Throughput: Thermodynamics and the Economy

Commercial firms, governments, households, and all other institutions that create wealth are, of course, subject to the "rules of the game" described in Chapter 1. As noted earlier in this chapter, the throughput model places the economy within its ecological and physical context and describes it in terms consistent with the first and second laws of thermodynamics. In a common use of terms, production is seen as the creation of material goods, which are then "used up" in providing a flow of services to producers or final consumers. This is an error that may be of little concern in the analysis of most economic problems, but for the economics of the environment and sustainable growth, the economy must be seen within the context of the ecological system and the governing laws of thermodynamics.

This context is represented by the throughput model shown in Figure 6.4. Low-entropy resources (A) are drawn from the environment and transformed into consumer services (B) with the accompanying production residuals (C) that are the embodiment of high-entropy energy. In a similar manner, consumption transforms B into high-entropy residuals (D). C and D are, of course, returned to the environment.

Production sector: A = B + C

Consumption sector: B = D

Economy: A = C + D

The economy, which sustains and enhances human life, is seen as an energy transformation process—no different in principle than the "economics" of other species that exist on energy transformations within their ecosystems. What we take from the environment we ultimately put back as residuals, which often damage the environment. Nothing is "used up" in this context of our habitat as a closed system.

A casual interpretation of Figure 6.4 suggests that economics is the study of the process of continuously burying ourselves in our own garbage—a characterization that would delight the most vociferous critics of economics but not quite

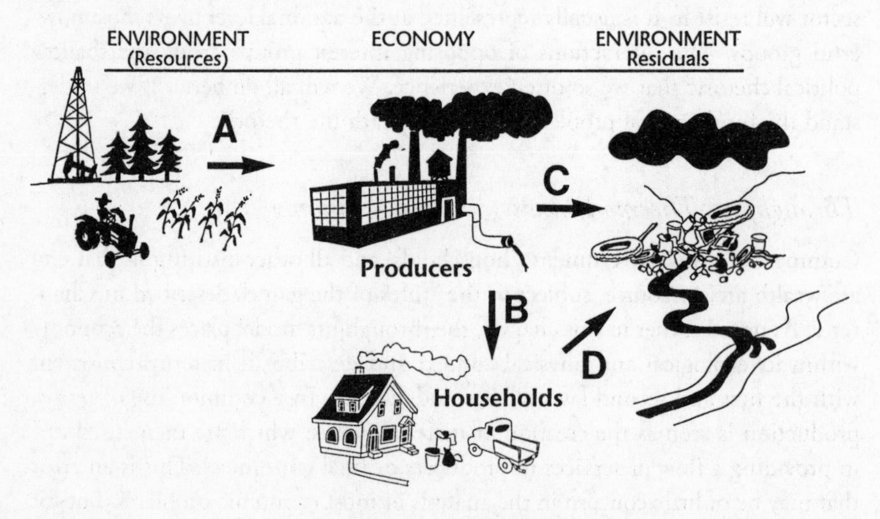

Fig. 6.4 The Throughput Model.

what we have in mind. Although production and consumption are entropy-related processes, the earth is not a closed system with respect to thermodynamic laws. Like all species, we are sustained and replenished by the sun. Entropic collapse is avoided or, at least, postponed.

The throughput of energy and material is essential for our existence, but our earlier discussion of the fact that wealth is not a simple question of material goods informs us that there is no one-to-one relation between the energy and material stocks that support this throughput and the flow of income and consumption services. The production challenge of sustaining our existence is to continuously increase the *ratio* of services provided to the throughput of energy and material resources. This ratio will change in response to energy, material, and environmental efficiencies and to the substitution of less scarce resources for threatened resources. Recycling takes place within the economy and, if effective, reduces the resources required to produce a given level of consumption.

All of these changes are amenable to, and require a focused change in, technology. If consumption is to continue growing toward some level that indefinitely sustains the world's population at a satisfactory standard of living, the ratio of wealth and income to the inputs of many exhaustible natural resources will need to increase continuously. Efficiency and recycling are important in this process. There are some exhaustible resources that are essential for life, and significant material resources will always be required. Substitutions for these resources are limited, and the energy must be available for continuous recycling. However, the core mechanism is the substitution of renewable for exhaustible resources. Technical progress is not a magic bullet that guarantees our future—as claimed by the more extreme optimists. However, it is the *sine qua non* of energy and material sustainability at anything above a basic level of consumption for a relatively small population.

The policies that support these changes need to be critically evaluated in terms of their energy and environmental impacts. Specific policies and consequent projects on recycling and resource substitution sometimes have been found to have adverse effects on throughput and the environment. "Recycling at any cost" is an oxymoron if the goal is to save resources and reduce throughput. The importance of energy is illustrated in the near certainty of our growing dependence on recycling for many of our material needs. Recycling, like everything else, requires energy conversion, and energy is not recyclable. In formulating policy, a broad perspective on thermodynamics and economic interdependencies is required to avoid unintended consequences and simple waste.

Box 6.1. Recycling and Other Conservation Measures

In 1990, the Environmental Defense Fund, in the interest of what it believed to be more environmentally friendly packaging, induced McDonald's to (1) abandon a polystyrene recycling program it was planning to take nationwide and (2) substitute compound plastic-paper wrapping for polystyrene in its packaging. "Life Cycle" studies of the processes involved have claimed that the overall environmental effect of this action may have been negative and that polystrene and other plastic packaging is relatively benign when full life cycle effects are considered. To resolve this question, we would need to compare such things as the recycling potential for compound paper products versus polystyrene, the environmental impacts of the production processes, and the energy transformation involved in the production and transportation of the alternative packaging materials. Advocacy for a cleaner environment and sustainable energy does not necessarily result in advancement toward either goal. The requirements for effective recycling are the same as for any other economic process—*efficiency.* In broad form, the overall use of energy, environmental, and natural resources must be reduced.

If the associated pollution problems can be resolved, the best use of waste paper may be as fuel for electricity generation rather than for recycled paper products. This substitutes a renewable resource for fossil fuels. Since power plants are ubiquitous, transportation costs would be low. Separation requirements (a largely hidden cost) would be reduced since virtually all paper products would be acceptable. Experiments using *unseparated* household trash as fuel for electric generation have been run. Although a high proportion of household waste is combustible, the resulting boiler corrosion and pollution problems from unseparated waste were not acceptable.

The development of battery-driven electric automobiles has received government subsidies as a potentially efficient and "pollution-free" substitute for the internal combustion engine. Electric motors have energy efficiencies in excess of 90 percent in many applications, with very little residual pollution. However, the process from fossil fuel extraction, electricity generation and transmission, and charging and discharging batteries is subject to a variety of efficiency losses and pollution problems. The new hybrid cars now available from Honda and Toyota, while relying on the internal combustion engine, may prove both cleaner and more efficient when full energy cycles are compared. In addition, given the current and expected near-term state of battery technology, they are likely to provide superior service for most consumers. Of course, future developments such as fuel cells (also the recipient of government subsidies) may trump all of the currently available mechanisms.

Exhaustible Resources: Markets, Discounting, and the Future Use of Specific Resources

Without enforceable property rights to the environment as a waste receptor, we know that markets cannot effectively solve most environmental pollution problems. At the other end of the throughput process that connects an economy with the natural world of thermodynamics, there is the question of dividing low-entropy energy and material resources between present and future generations. This requires a comparison of present and future values for these resources. Economists do this all the time, and it involves an understanding of interest rates as they represent the productivity of resources and the time-cost of waiting. The presentation here is very straightforward, but it requires concentration on some details.

How do markets take account of future needs for exhaustible resources that otherwise might be used today? We can address this question with a hypothetical example of a specific resource. The *net price* of a barrel of oil in the ground is the market price less all costs of extracting it and moving it to the market. Suppose that this value is \$1. Should the owner of the oil well pump and sell the oil? The answer depends on the costs and benefits of *not* pumping the oil. The cost is the lost revenue of \$1, and the benefit is the expected net price sometime in the future. The interest rate is crucial to this comparison. Suppose the relevant inflation adjusted interest rate is 5 percent. The owner can pump the oil, put the money "in the bank," and have \$1.05 at the end of one year. However, if the expected net price of oil is \$1.05 *or more* at the end of one year, the owner can do as well or better by leaving the oil in the ground. If the oil is to be left in the ground for fifty years, the expected net price fifty years from now would have to be at least $\$1(1 + .05)^{50} = \$11.47$ after allowing for any inflation.

Resource owners look at many factors, including present and expected future prices, when making decisions on extraction. If they do not expect prices to be sufficiently high at future dates, then more oil will be pumped today. This lowers its present price and decreases reserves, thus changing the relationship between present and expected future net prices—with the result that postponing extraction becomes more attractive. In this process, involving the supply and demand for resources both now and in the future, a continuous link between present and future prices is established. Thus, future demands for oil and all other depletable resources are *discounted*—that is, at an interest rate of 5 percent, an expected net price of \$11.47 fifty years in the future is discounted to a value of \$1.00 as seen by today's generation. It is what \$1.00 put in the bank today would yield if left for fifty years. A belief that future generations will pay a higher price for resources induces the present generation to conserve resources for their use.

Table 6.1. Relation of Present and Future Net Prices

		Future Expected Net Price Required to Induce Conservation for:	
Today's Net Price	Rate Interest	One Year	Fifty Years
$1	3%	$1.03	$4.35
$1	5%	$1.05	$11.47
$1	10%	$1.10	$117.39

The *discount rate* (which is the interest rate used by the resource owner in this context) is crucial to the comparison. If it is 10 percent, then an expected net price of $1.10 is required to conserve the oil for one year. A price of $117.39 is required to conserve it for fifty years. If the discount rate is 3 percent, $1.03 is required at one year and $4.39 is required at fifty years. Thus, given a particular expected net price in fifty years, the higher the discount rate, the more likely is the decision to extract an extra barrel of oil today rather than later. High interest rates require high expected future net prices to induce the postponement of extraction. As summarized in Table 6.1, the power of compound interest becomes obvious in the comparison of the effects of 3, 5, and 10 percent interest rates on the future net price bids required to preserve resources for future use.

The implications of the relationship between present and future generations in the allocation of exhaustible resources were first developed by Harold Hotelling in 1931.[8] The major theorem is that market interactions between buyers and sellers will result in the eventual continual increase in the net price of exhaustible natural resources at the rate of interest. The process will continue until the resource is depleted. The term *eventual* is intended to capture some of the nuances of this process as the result of a myriad of events, such as discovery, changes in technology and consumption patterns, and the development of substitute resources. All of these potential and actual events create great uncertainty in resource markets. They disrupt the smooth flow of the link between present and future resource prices. Natural resource economists note that, as a consequence, natural resource owners use discount rates that are too high relative to the appropriate long-run social discount rate covering all natural resources. In other words, relative to the social optimum, they value the security of present income too highly in the face of uncertainty about the future. This means that current prices of natural resources are too low—a market failure resulting in too little conservation for future generations.

The evolution of resource pricing over time was not a major economic issue of the 1930s. Hotelling's path-breaking work received little attention until the 1970s, when the OPEC oil crisis and the need to respond to the *Limits to Growth* models generated serious interest in natural resource economics and the dynamics of long-term growth with exhaustible resources.

Much of the work resulting from this renewed interest was in the development of a theoretical base. However, a very informative and insightful article produced by the economic historian Nathan Rosenberg will help us put some meat on the theoretical bones.[9] Rosenberg went in search of historical price series that might conform to Hotelling's theory.

Perhaps surprisingly, the best example he found was timber, a long-cycle renewable resource. He tells a fascinating story of the low value of wood in colonial America and the response to the increasing value of stumpage (net price) over time. Timber was superabundant in most of colonial America, while labor and capital were relatively scarce and costly. Trees were often a by-product to be eliminated in the clearing of agricultural land. Thus, particularly on the frontier, we observe an intense substitution of wood for other factors of production—log cabins, split-rail fences, and even roads laid out with split logs. European wood product manufacturers came to observe the wonderful woodworking machinery developed in the colonies. The most famous example is the gun stocking lathe that would turn out a gun stock from a single piece of wood. They went home in despair, noting that the processes were so terribly wasteful of wood—a scarce commodity in Europe. Bridges and the tremendous railroad trestles of the U.S. West were produced from wood long after Europe had turned to an emphasis on stone and iron.

As the net price of timber rose through the nineteenth century and particularly in the twentieth century, Rosenberg's story turns to one of adaptation to the increasing scarcity of timber. Construction and other applications of veneers, plywood, concrete, steel, composition boards, aluminum, and vinyl are all, at least in part, a response to the increasing net price of timber. This has also contributed to the development of professional forestry aimed at maximizing forest productivity. Another interesting indication of the increased value of wood is that one no longer sees mountains of sawdust at sawmills. Virtually all scrap is either used with adhesives to form wood products or used for fuel.

The economy's engines of adaptation to growth with exhaustible resources that we will continue to emphasize are *substitution* and *technological change.* Economists often tend to think of these as distinct concepts since they are separable in the mathematics and geometry of production theory. But Rosenberg observes that, in the real world of adaptation, they are inseparable Siamese twins. Only small portions of the possible relationships between inputs of resources and outputs of goods and services are known at any one time. The process of substituting low-cost resources for high-cost resources in producing an output is always one that requires discovery and technical change. The fuel that drives these engines—that guides substitution and technical change in desirable directions—is the rising relative price of the scarcest resources. In a nutshell, "the squeaky wheel gets the grease."

Energy, Sustainability, and the Zero Discount of Future Human Well-Being

It is often argued by environmental activists that future generations have as much right to consume from the existing stock of exhaustible resources as the present generation. This implies that the appropriate discount rate for future net price bids should be zero. Future generations should pay the same prices that we do (after allowing for any general inflation) rather than the higher real prices demanded by the self-interest of the present generation. This simply is not feasible for any single resource—even if we knew the actual size of the existing stock. If each of the virtual infinity of future generations has an equal right to the finite stock of a resource, then the allowable consumption for any single generation must be essentially zero. Clearly, this would not allow for the use of a resource to meet present human needs. The very act of allocating the use of a single exhaustible resource to the present generation, implicitly or explicitly, discounts the future use of that same resource at some positive rate of interest. The exception to this rule is the case in which a resource is expected by the present generation to become useless before the known reserves are exhausted.

Our ancestors discounted our demands for natural resources—seemingly to the point of ignoring them altogether—yet, in the developed world, we are markedly wealthier than they were. How did this happen? The answer is straightforward! Up to the present, they have bequeathed to us in the form of social institutions, and skills and technology embodied in reproducible capital, much more than they took from us in the form of access to depletable and renewable natural resources.

Essentially, as part of our inheritance from the past, *reproducible capital* has been substituted for *natural resources.* Although our needs for specific resources have been heavily discounted by our ancestors' activities, our overall need for income-producing wealth has not. In the developed and developing world, per capita incomes and consumption (again, broadly defined) continue to rise.

Can this process continue to provide our heirs, into the indefinite future, with the energy required for a worldwide level of economic and social opportunities that are at least as rich and diverse as our own? The future uses of particular exhaustible resources are necessarily discounted and, even with recycling, many of these resources eventually may be depleted. It does not follow that the future use of *all* resources must be discounted at a positive rate of interest. The "equal rights of the future" activists are absolutely correct when seen within this broader framework. The basis of sustainability is, in fact, the zero discount (actually negative—with growth) of future needs for a broadly defined package of income-producing wealth. This means that we must leave for future generations a package of resources—inclusive of such important components as knowledge and the quality of the environment—that is *at least* equivalent in producing human well-being to what we have received from previous generations.

We have documented a large and growing population and the high and growing use of fossil fuels and other natural resources, with associated environmental problems. With the limits imposed by finite stocks of specific energy resources and a vulnerable environment, we conclude that it may be very difficult to do for our heirs what our ancestors have done for us. The process of providing for the future will necessarily be the same, but there is no guarantee of success. Given this understanding, the policy question becomes: what can we prudently do within this generation to increase the likelihood of an adequate resource base for future generations?

Sustainability requires a volume and structure of energy transformations necessary to provide a continuously adequate combination of capital stocks (wealth) composed of reproducible physical, human, biological, and environmental capital and exhaustible natural resources. This inevitably will involve the substitution of technology—embodied in natural, human, and nonhuman reproducible capital—for exhaustible resources. Ever-decreasing stocks of exhaustible resources must be replaced by other forms of capital.

Because of the large amount of diversity and uncertainty associated with implementation, policy is best viewed as a continuous *process* of decentralized adaptation rather than as a goal or a set of rules. Markets in conjunction with other economic and social institutions must establish relative social values for all resources across the present generation, as well as between present and future generations, in order to guide efficient and equitable resource use. Policy should be flexible and adaptable so as to:

- adjust to outcomes and changes that will continually surprise us
- reduce the economic and social waste of policy-generated disruptions.

Growth may be possible on a very long term basis. But this is only true if growth is correctly perceived in terms of meeting human needs and wants rather than as the indefinitely increasing consumption of material resources. To increase and sustain human well-being, a necessary condition is to increase and sustain *consumption* (services flows) stemming from our stock of *wealth.* The components of consumption that meet basic and acquired human needs will continuously change with the changes in, and interactions between, available resources and our social, cultural, and economic institutions.

Policy: Conserving Energy Resources and the Environment

The goal of an energy policy is to begin a process that effectively values the environment and the energy needs of future generations. It is presumed that this process takes place within a democratic framework that recognizes both diverse cultures at home and abroad and the difficulty of establishing political motivation on behalf of policies where the benefits are both remote and uncertain.

Recent worldwide trends in environmental policy show a growing concern for incorporating the incentive structure and interdependence that are the hallmarks of the marketplace. The basic policy outline provided here follows this trend. Concerns are in achieving a balance of the following:

- effectiveness in meeting policy goals
- efficiency in the avoidance of resource waste and excessive social and economic disruption
- equity in protecting those with the lowest incomes from economic loss.

Whatever the source and nature of the motivation that raises our concern for the environment and the energy needs of future generations, it will eventually enter the policy process and be formulated into law—and law, even in a relatively free and democratic society, is ultimately coercive. It is a question not of *whether* the policies will impose new exclusions (property rights) on our use of resources but of *how* these exclusions will be imposed.

Supplementary Charges

The policy perspective proposed here is the use of supplementary charges as instruments of public involvement in the marketplace for the purpose of conserving natural-environmental resources. The relative success of this approach compared to direct regulation, and the growing understanding of its underlying mechanism of economic incentives, has led to its support by virtually all major environmental groups as an important concept in the policy tool kit. The underlying economic principle is that, for an efficient allocation of resources used in production, each resource (including the environment) must carry a price that reflects its true social value. Underpriced resources will be overused—that is, "wasted." When market forces fail to achieve this *full social cost pricing,* socially desirable resource use requires a policy supplement. We can limit use by rationing (regulation) or by supplementary charges that will affect pricing (for example, resource taxes, pollution taxes, or marketable permits). We will continue to do both. However, in policy debates, supplementary charges have grown in prominence as their effectiveness has been demonstrated in a large number of resource management cases. They have the characteristics required for a productive basic policy. The power of supplementary charges lies in their mimicking the market in the sense of creating incentives for decentralized, productive decision making that conserves social resources in a complex, interdependent economic system. Supplementary charges are not a solution to be mechanically applied to all resource and environmental problems. Conservation inevitably will require other forms of regulation and social change in a complex policy milieu.

Energy and pollution charges: The regulating authority determines a charge in the form of a tax for each unit of energy used or pollution emitted. A carbon tax can be viewed as either a carbon emissions tax or a fossil fuel tax. Any unit of carbon emissions that can be eliminated at a cost that is less than the tax will not be emitted. The regulator seeks a charge level that will achieve the targeted abatement. Those who are unable to substantially reduce emissions at low cost are disadvantaged in the marketplace by the necessity of paying the charge. This creates a continuous incentive to seek low-cost methods of conservation that is identical to all of the other cost pressures that operate on producers and lead them to conserve costly resources.

Marketable permits: The regulating authority auctions a quantity of permits that represents allowable energy use or emissions. The auction price is determined by, and measures the cost of avoiding the last unit of energy use or allowable emissions. Those with less-efficient processes and higher costs must pay for the permits and suffer a competitive disadvantage. As with taxes, this creates incentives to innovate in energy use and pollution control. Permits may be resold, and their changing price will reflect progress (or the lack of it) in the discovery of low-cost abatement techniques.

In principle, energy and environmental taxes or permits can be designed to yield exactly the same outcomes. However, in the real world of policy, this would be almost impossible to achieve. The following presentation will concentrate on taxes for simplicity and because there is generally less disruption and more adaptability to fixed price increases than to fixed quantities. However, permits are a viable and, in some cases, a superior alternative for technical or political reasons.

Supplementary charges were proposed for environmental policy more than thirty years ago. For very different reasons, they were opposed by both environmental groups and the business community. Environmental groups saw the policy as selling our priceless heritage like an ordinary commodity and regarded market-related activity as *the* problem rather than as a potential solution. The business community knew that the policy would work, that the burden of reducing commercial pollution would fall squarely and immediately on them as intermediaries in the production-consumption process. They preferred to take their chances with the traditional regulatory process—perhaps believing that the environmental fervor would go away fairly quickly. One of our best hopes for a viable sustainable energy policy is in the clear evidence that environmental concerns have not declined.

Environmental and business groups both have substantially reversed their former opposition to supplementary charges. Environmental groups have observed the effectiveness of this approach in a variety of circumstances. Business groups have experienced the excessive costs of traditional regulation, costs that are ultimately paid by their customers and by all of society.

Although this policy concentration may seem fairly narrow, we shall see that its effects are very broad. Also, it reinforces rather than contradicts the cultural and value issues addressed by Richard Wilk in Chapter 4. We will attempt to place policy recommendations in the context of these broader issues and will also address the difficult and related issue of policy formation in a democratic society that places a high value on personal freedom—the subject of Randall Baker's contribution in Chapter 5. These contributions provide a multifaceted perspective on the problems of allocating community resources, in which many of the benefits will be enjoyed by strangers who are remote in both time and place.

There are many ways in which underpricing sabotages the conservation goal or, alternatively, many ways in which full social cost pricing supports it. The benefits of full-cost pricing fall under two major categories:

- broadly and efficiently conserving threatened resources
- broadly pushing technology and investment toward energy and resource conservation.

There is near-uniform agreement among economists who study exhaustible resource problems that markets do not *adequately* take into account the needs (bids or claims) of future generations for resources. The outcome is that these resources are underpriced, and therefore overused, in today's markets. Uncertainty with respect to both the supplies and the needs for resources in the future inevitably leads to shortsightedness by exhaustible resource owners. Uncertainty is reduced for society as a whole since there is diversification over many resources. Unlike the single resource owner, all of our collective eggs are not in one basket. This disparity in uncertainty leads to a market discount rate used by

single-resource decision makers (who are subject to cor
markedly higher than the relatively risk-free social discou
priate to induce an efficient postponement of extraction. I
fession, the "discounting" of future needs for resources is e
earlier, because of compounding, a small increase in the discount rate causes a very large increase in expected price bids from future generations that are required to induce a postponement of extraction. This volume has appropriately concentrated on energy resources, since everything else depends on energy conversion. However, the analysis applies to all exhaustible resources.

A tax on energy throughput would help correct for the shortsightedness of markets with respect to future energy resource needs. Raising current prices of these resources would lead to a decrease in the quantities demanded and the greater availability of these ultimately exhaustible resources in the future. On the demand side. the price of *every* good and service will be affected in proportion to the energy used in its production and distribution. We have, from the OPEC experience, such substitutes for high-priced energy as sweaters, insulation, insulated glass, higher-efficiency furnaces and automotive engines, lower thermostat settings, car pooling, bicycles, and public transportation. Many of these substitutions have been reversed with the falling real price of energy in the 1980s and 1990s. Current tax policy in the form of depletion allowances is actually perverse in that it *subsidizes* extraction.

The overall effect of resource and effluent taxes in raising the price of fossil fuels also has a vital impact on energy supplies. If the price of fossil fuels reflects their true social cost—including conservation and environmental costs—then relatively clean renewable energy resources will become more competitive. This will stimulate research to develop the technology and lower the cost of renewable energy resources. Compare this to subsidizing the development and marketing of alternative energy sources so that they can compete with underpriced fossil fuels. An all-too-typical example is the ethanol program. The subsidy now seems to have the status of a political entitlement for farmers and a few very large agricultural processing corporations. It has had no conservational impact on the demand for energy since it did not significantly affect the price of energy.

Markets, left to themselves, will fail to conserve most environmental resources, because they are *common property resources* and are not traded in markets. Indeed, an unregulated competitive market would be more or less forced to exploit these resources because they are regarded as "free" or "unpriced." We buy our goods at artificially low prices, but the resulting *social* cost of this activity is pollution and energy depletion. This social cost of damage to the environment and future generations can be very high—some would say life threatening.

The traditional policy approach to environmental pollution has been

…le 6.2. Ratio of Command and Control to Least Cost Pollution Control

AIR POLLUTION	ratio: CAC/Least Cost
1. Particulate Control: Lower Delaware Valley	22/1
2. Hydrocarbon Control: All Domestic DuPont Plants	4.15/1
WATER POLLUTION	
1. Biochemical Oxygen Demand: Delaware Estuary	3.31/1
2. Biochemical Oxygen Demand: Oregon, Willamette River	1.12/1

Source: Tom Tietenberg, *Environmental and Natural Resource Economics* (New York: Harper Collins, 1996).

through specific and detailed regulation—the command-and-control (CAC) approach. However, a large number of studies have found that this approach is nearly always more costly than the pricing approaches (supplementary charges) that provide the structure for least-cost outcomes. The cost advantage of the pricing approach can be small, but it is often very substantial. The resource cost of CAC can be as much as twenty times the cost of controlling the same amount of pollution through a pricing approach.[10] Table 6.2 illustrates these points with some examples from research.

The supplementary charge approach to policy has important consequences:

- A given amount of pollution control can be achieved through supplementary charges for a fraction of the cost to consumers, and of throughput to the environment, when compared to the CAC approach that currently dominates policy. This follows primarily from the fact that a supplementary charge will concentrate the required abatement at the lowest-cost sources—both within a facility and between facilities. Facilities with high abatement costs will initially pay the tax, while facilities with low abatement costs will reduce pollution. CAC regulation can never accomplish this since the regulator can never know enough to fairly and efficiently allocate the required abatement. Effluent charges have an inherent fairness component because those who do the least cleanup pay the heaviest charges—and are at a competitive disadvantage.
- Perhaps even more important, effluent taxes establish a *continuous incentive* for producers to innovate and find more effective ways to conserve energy and control pollution. The incentive is to escape the charges by energy conservation and additional pollution control at a low cost. Conversely, the record of CAC in this regard is often poor and sometimes adverse. The incentive for the polluter is to meet the legally enforced pollution standard and then turn attention to other management and technical problems. It is often the regulator who dictates the technology and, in any case, to innovate is to risk having stricter standards imposed by the regulating authority.

- Perhaps most important, economic costs translate into political costs. Over time, lowering the cost of pursuing social goals has a powerful effect on our willingness to pursue them.

As noted earlier, pricing approaches are gradually supplementing or replacing other forms of regulation with the support of most major environmental groups. Also, as noted, direct regulation will continue. An example would be in areas where it is difficult and costly to monitor compliance. It is, therefore, important to continue with other innovations in the regulatory field.

Ordinary taxes are justified by an important social purpose, but they have a damaging side effect. They fund the public sector, but they distort the incentive structure of economic decision making and reduce the efficiency of the economy in terms of its ability to provide for the best use of resources. In contrast, pollution and energy taxes, when seen as partially replacing ordinary taxes, achieve three important social purposes:

- Like ordinary taxes, they fund the public sector.
- They *improve* the efficiency of the economy—especially in efforts to control pollution and reduce the throughput of energy and material.
- They help to conserve exhaustible resources while smoothing the transition from present to future energy technologies.

Policy Implementation

In Chapter 5, Randall Baker has provided us with a valuable perspective on the policy process. The particular difficulty of policy formation in response to long-term problems is a well-established phenomenon. The resulting frustration has led some commentators to state implicitly or explicitly that resolving the sustainability issue will ultimately require a substantial reduction in traditional freedoms. It is a view that is expressed by Robert Hielbroner, as quoted by Norman Care in Chapter 7. The perspective in its extreme form is that we can have democracy and personal freedom or we can have sustainability—but we cannot have both.

This is a devastating view of our future choices that I do not share, but it does drive home Baker's point about the difficulties we face in policy formation and implementation. These difficulties are not insurmountable, as the serious political debate on the long-term viability of the social security system demonstrates. However, they do require that we take the limitations of the policy process in a democratic society very seriously.

Baker's conclusion that we will probably need to concentrate on the environmental dimension of the energy problem is less limiting than it might seem. Effective interventions into the throughput process inevitably impact both the low-entropy energy and the high-entropy residual ends of the spectrum. Policies can be designed to maximize the effect of intervention. For instance, substantial carbon taxes in response to suspected global warming from greenhouse gases are equivalent to fossil energy taxes. The same is true for other environmental tax or marketable permit systems.

Ease of transition plays a prominent role in the following discussion of policy implementation. The topics covered include taking account of our ignorance, gradually implementing and indexing energy taxes, substituting energy taxes for other forms of taxation, and considering equity and fairness. Reducing the inevitable disruption caused by policy implementation cannot eliminate resistance and opposition to the policy, but it can reduce it. This lowers the level of the required crisis atmosphere and the inevitable rhetoric of disinformation. This is a good thing. Panic rarely leads to effective decision making.

Ignorance

Randall Baker provides us with a cautionary tale in Chapter 5 when he describes the foundering of previous prognosticators of gloom and doom. Can Micawber's technological fix save us from our "profligacy" into the indefinite future? Possibly, along with other social changes, such as a stable population. But we cannot know that it will. Could Cassandra's call for stringent conservation of existing resources save all future generations? Not by itself. As noted earlier, a finite resource divided by all future generations allows zero consumption by any single generation. The greatest probability of success lies in policies that combine technology, social change, and conservation and that do not pretend that we know more than we do.

William Stanley Jevons, a prominent economist (Cassandra) of the second half of the nineteenth century, was a leader of the coal conservation movement in Great Britain. Living near the beginning of the world's most dramatic sustained growth of per capita income, he correctly perceived that the Industrial Revolution was based on mechanical energy. He observed that the use of such energy would continue to grow. Looking at proven and estimated reserves of coal, he saw cause for grave concern about our prosperity in the twentieth century and on into the future.[11] Suppose the concerns of the coal conservationists had resulted in stringent conservation of this fuel. Would we be grateful? It seems unlikely. Growth in per capita income probably would have been slower. We cannot know how much slower since we don't know what the technological response to coal stringency would have been. However, it is a near certainty that

Box 6.2. The Industrial Revolution, Affluence, and Consumerism

One of the most dramatic illustrations of the power to change human lives exerted by that nineteenth-century phenomenon we know as the Industrial Revolution is provided by Stephen Ambrose in *Undaunted Courage.* Ambrose makes a simple statement that the transportation technology available to Lewis and Clark was much closer to that of the Roman Empire than to that available to most areas of the world a mere one hundred years later. Affluence for small, elite groups was well known prior to the rapid innovations in the use of mechanical energy and the spread of specialization and market transactions that characterized the Industrial Revolution. However, it was the Industrial Revolution that spread this affluence to larger and larger segments of the population—by providing sustained and unprecedented increases in per capita productivity and incomes beginning in the late eighteenth century. Following Richard R. Wilk (in Chapter 4), we may characterize affluence as the *opportunity* to translate "wants into needs." Affluence seems to have acquired a bad reputation in direct proportion to the number of people who are able to enjoy it. The major concern, of course, is with the energy and other resource transformations that affluence requires for its sustenance with current technology and capital structures. This and related social and ethical concerns are captured with the pejorative label *consumerism.*

We need to address consumerism with some care. The problem is that the term tends to be applied to someone else's consumption. I am sitting in front of a computer that uses relatively little energy but that is much more powerful than is "needed" for word processing or virtually anything else that I do with it. I simply accomplish various tasks more quickly. I am in a house that far exceeds my "basic needs." The energy used for a vacation trip to Europe, or for an academic to deliver a paper at an international conference, would run a sport utility vehicle (SUV) vigorously for an entire summer of camping. If we outlaw SUVs as "rampant consumerism" and the former camper acquires a taste for European vacations or water skiing, where is the conservation? How do we deal with the enhanced public health and longevity that are associated with affluence? How do we address the fact that the lowest 20 percent in the U.S. income distribution live at levels that would be middle class or above for the bulk of the world's population? It is not just difficult to throw out the bathwater while retaining the baby; it is also surprisingly difficult, and politically divisive, to distinguish between them. The issues here are the complex interdependence of an economic system and the "law of unintended consequences."

(*continues*)

Box 6.2. Continued

Condemning particular forms of consumption may not be the best use of limited political capital in our efforts to establish sustainable energy policies. This is especially true if the primary outcome is analogous to pushing in on one side of a balloon and watching with surprise and dismay as it bulges out on the other side. If we are to increase per capita incomes for poor nations and people, and to maintain income levels beyond the foreseeable future, we will need approaches to reshaping consumption patterns that are much more comprehensive than the searching for scapegoats among an affluent population or the condemning of particular forms of consumption.

oil, as a close substitute for coal, would have been exploited much more quickly than it was. This is a resource that, relative to our presently perceived needs, we know to be in much shorter supply than coal. Do we know more than the coal conservationists? We do—but we need to recognize that our ignorance will always outweigh our knowledge. It is very difficult to predict what future generations will most want from us.

Ignorance is not a good thing, but *recognizing* ignorance can help us avoid serious errors. An earlier section of this chapter noted the error of concentrating directly on *goals* rather than *processes* in policy formation and also noted the similarity between ecological and economic systems. Both appear to have considerable stability within certain bounds, but when their complex balance is disturbed by interventions, they can generate unexpected and destructive outcomes. These are the unintended consequences of economic policy. The size of a serious energy intervention carries the risk of substantial, unpredictable, and wasteful disruption.

Gradual Implementation and Indexing of Energy Taxes

What is the appropriate level of energy taxes to achieve optimal conservation? Research has given us considerable insight into some dimensions of the problem but, fundamentally, we do not know. In light of the preceding discussion, this is not particularly debilitating. The observation applies with equal or greater force to any other regulatory policy we might follow. The level would be established on the basis of limited knowledge and the political process. Any substantial tax would set in motion the dynamic process of conservation and would also provide the basis for future modification and supplementation. Political limitations make it much more likely that the tax will be too small rather than too large.

OPEC has taught us that the sudden imposition of a large energy tax would be extremely damaging. A substantial tax needs to be imposed gradually. In this case, good economics and good politics go hand in hand. Imposing a small tax that increases automatically by preestablished increments over a period as long as twenty years has many advantages. Politically, it would allow for the imposition of a larger tax. It allows time for adaptation and, therefore, markedly reduces the disruption of consumers, workers, and existing capital structures. Plans can be made with the knowledge that energy costs will be substantially higher in the future. For example, the current owner of a sport utility vehicle would not be blindsided by the prospect of a sudden large increase in operating costs and would be less likely to be galvanized into opposition with widespread support. Over time, however, we would expect the sales of SUVs to decline substantially as "needs" were reassessed in contemplation of high fuel prices. Vehicle production would adapt accordingly. Similar reassessments of need, and adaptation over time, would take place with respect to the efficiency of space needs, energy-consuming production processes, other modes of transportation, and so forth. The list is endless, and the effects would be comprehensive. By minimizing disruptions, gradual adaptations over a substantial period may well allow real incomes to continue rising. The economic importance of continued growth for developing countries and the political importance for all countries should be obvious.

Since the policy goal is to raise the real price of energy over time and to maintain it at a high level, the energy tax would need to be indexed for inflation. The current dollar price of a gallon of gasoline is near its peak level of the late 1970s. However, the *real* price (the price relative to the prices of other goods and services) is near historic lows. The brief decline of energy consumption in the 1970s and the subsequent growth below historic trends were occasioned by gasoline's high relative price. When inflation eroded the high relative price, the conservation impact lost its momentum. Indexing means that the energy tax automatically increases with inflation.

Substituting Energy-Effluent Taxes for Other Forms of Taxation

In our local community, a successful trash recycling program was implemented in the mid-1990s. It consisted of increasing the convenience of recycling by offering pickup, coupled with a $1 charge per bag of trash that was not separated for recycling. It is an example of how even modest charges can have important conservation effects. Over 30 percent of household waste is now recycled, and the program is still expanding. However, despite this very modest charge for trash pickup services, the opposition rallied around the position that the program was just another round of tax increases by a greedy government—disguised as a conservation program. We are all familiar with similar views expressed in response to proposed energy taxes.

Any proposed tax increase is going to create political opposition. However, a carefully constructed policy will seek to soften the blow. In this light, the issue of sustainability needs to be clearly separated from the issue of the overall size of government. Similar to gradual implementation, this is a means to reduce political opposition and the required level of "crisis mentality" needed to generate political action. Explicitly substituting energy taxes for other forms of taxation is a direct way of accomplishing this goal. It is a move that frees the advocate from the suspicion of a hidden agenda and thus helps to blunt the organized opposition. It also removes a potential constituency for the organized opposition by reassuring the ordinary taxpayer. While taxpayers are faced with a price structure that more accurately reflects the true social cost of energy use, their after-tax income is not directly reduced. In this way, a larger energy tax can be imposed than would otherwise be politically feasible.

In April 1999, Germany initiated an "ecological tax reform" that incorporates both of the features that we have proposed. An energy tax was gradually imposed over a four-year period (2000–2003). It is a partial substitute for their version of the social security tax.

Equity

"Where will the burden fall?" "The burden on low-income families will be excessive!" This is a serious economic and political problem. For the policy to be effective, everyone—both "rich" and "poor"— must conserve. By most definitions, there simply aren't enough of the "rich" to provide the basis for an effective long-term policy. Everyone must contemplate conservation possibilities in the context of high energy prices—yet we do not want the policy to increase poverty.

A well-known way to respond to this dilemma is to provide a refundable income tax credit that is equivalent to the energy tax for some "standard" quantity of per capita energy consumption. The result is that the income is available to consume energy up to the standard at higher prices. However, the relatively high price on every unit of energy consumption continuously induces attractive substitutions as they become available to the household. There are other issues of tax equity in the substitution of energy taxes for other taxes, such as the relative regressivity of energy taxes compared to income taxes. This means that, in comparison with income taxes, energy taxes will be a higher proportion of low incomes than of high incomes. These issues will need to be addressed in the structure of offsetting tax reductions. Substitution for the social security tax is a good example, since this tax on wage income is also regressive.

The same problems of rich versus poor emerge on the international scene in connection with the global warming issue. They are clearly seen in the efforts to find acceptable compromises between developed and developing nations on the policy proposals for carbon restrictions and marketable carbon permits contained in the Kyoto Protocol of 1997. These are the most comprehensive proposals for supplementary charges that we have to date. The position that the developed nations, which use a disproportionate amount of the world's energy flow, should subsidize the less developed nations has strong resonance. The Kyoto Accords on global warming set the stage for moving the international community toward a sustainable energy policy. A problem with these accords is that they approach the issue of subsidizing the nations of the less developed world by exempting them from the policies that would create restricted fossil energy use for the developed world.

It is important to note that the policy of exemptions for developing countries from the stringent requirements imposed on the others is politically difficult and poses serious problems of unintended consequences. Developed nations are unlikely to provide and sustain a cost advantage to less developed countries in the production of energy-intensive goods—especially since less developed countries convert energy with less thermodynamic efficiency and control pollution less effectively than the developed ones. The shift in the production of energy-intensive goods from the developed to the less developed countries in the context of a global economy would likely be disruptive and would seriously compromise the goals of the program. Shifting carbon emissions from North America and Europe to other continents cannot conserve energy, reduce pollution, or slow global warming.

Of equal or greater importance is the fact that artificially cheap energy will encourage energy-intensive development, which will require damaging transformations of likely fragile economies with the nearly inevitable large future increases in energy costs. In the 1970s, the sudden, large OPEC-related increases in real energy prices were much more damaging to less developed countries than to the developed ones. In Chapter 2, John Sheffield stresses the importance of assisting development while energy prices are relatively low compared to their likely future levels, but it is not his intention to advocate subsidies that result in energy sales below full social cost as perceived by the present generation.

Consequently, we need to find ways to offset the costs of energy policies to developing countries that do not run counter to the primary goal. Subsidizing technological transfer and capital for energy-efficient production and consumption, population control, old-age security, and other social programs are examples from a long list of possibilities. This will not be an easy task, given the differences in cultural, political, and economic structures.

Political and economic considerations require developed countries to take account of the international scene in formulating our own internal policies. Energy issues must be addressed in the context of global competition. As an example, Germany found it necessary to exempt exported goods from their energy tax described earlier. A feasible, but difficult, approach is to apply tariffs to goods from countries that do not have similarly stringent energy policies. The size of the tariff or subsidy should be proportional to the energy impact of the traded good. Although tariffs are usually anathema to economists (because the real agenda is typically special interest protectionism), in this circumstance they give an exporting country the option of collecting the energy tax itself and using it to support domestic policy initiatives or of paying it to us. It can thus avoid

Box 6.3. Population Policy

We cannot prove that a sustainable energy future is unobtainable without the "artificial" constraints on population growth of a population policy. Neither can we prove that our future is assured if we follow a policy of severe constraints on population growth. What *is* clear is that a substantially larger world population at an acceptable standard of living makes it *much more difficult* to achieve a sustainable energy future.

Population issues are integral to Chapter 2. One of the most important relationships shown is the one between population growth and energy use per capita in developing countries (Figure 6.5). There are several factors underlying the relationship between economic development, per capita energy consumption, and decreasing birth rates:

- A relatively smaller agricultural population accompanies economic development. In some, but not all, cases it is less costly to house and nurture children in rural areas than in industrial or urban areas, and they are more productive as laborers.
- Higher education levels, especially for women, and job opportunities increase the opportunity cost (the economic sacrifice) of bearing and raising children.

- Alternative private and social forms of old-age security are developed, and parents become less dependent on children for old-age assistance.
- A relatively even distribution of income usually accompanies growth. This fosters a more rapid decline in population growth as the relative prosperity and its effects are spread over a larger segment of the population.

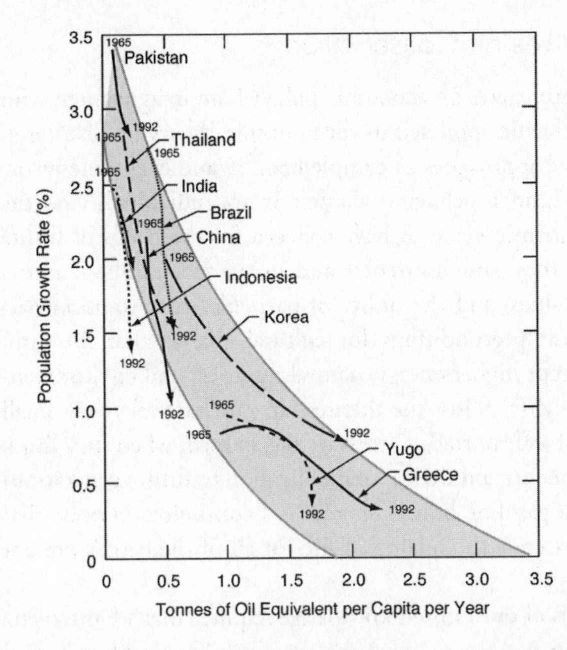

Fig. 6.5 Population growth and energy use per capita.

But there are many social-cultural and economic factors that strongly affect this process and the appropriate policies that could accelerate it. The policy goal should be to shift the population growth-energy use curve of Figure 6.5 to the left while enhancing the opportunity for per capita income growth—more like China and Thailand than Korea.

All three countries have emphasized education. Korea began development earlier on a heavy industry base, which explains much of its more substantial energy consumption. China has the most stringent birth control policies. Thailand started down the self-sufficiency, heavy-industry path but then shifted to an emphasis on agricultural productivity and light industry for export, which require less energy for a given value of output. These are but snippets from complicated stories, but they make the point that the population growth energy use curve is subject to policy interventions. A very important factor may be the decline in heavy industry as a preferred or even necessary engine of per capita income growth in the "postindustrial" era.

tariffs on its exports and subsidized imports as represented in the tax exemptions allowed by Germany. Environmental issues are now a standard feature in the debates on most trade bills and are being pushed forward as appropriate for international trade negotiations.

Conclusion: Perspectives on Conservation

Although my analysis concentrates on economic policy, I am in agreement with Richard Wilk on his multigenic approach to consumption theory in Chapter 4. Like Wilk, I see alternative approaches as complementary and even synergistic. High energy prices affect human behavior whether or not individuals view the environment as having intrinsic value or have concern for the needs of future generations. In addition, they raise awareness and foster debates about moral obligation, broad social values, and the utility of particular consumption patterns. These are important preconditions for cultural change that, in turn, increase willingness to accept higher energy, natural resource, and environmental costs. People turn off lights, adjust the thermostat, purchase relatively small cars or alternative forms of transportation, recycle, and so forth when they think about energy, the environment, and their moral obligation to future generations as well as their utility and gasoline bills. The goal is a continuous process that conserves resources and extends the quality of life for all of the earth's present and future inhabitants.

Proceeding on the basis of our limited knowledge requires that we intervene gently but persistently, do not try to solve the sustainability problem for all time, and move in directions that are consistent with our ability to create political will. What we can do is set the tone and direction for the two or three following generations that represent the limits of our ability to care deeply about the future. It is then up to the next generation, with their increased knowledge of the physical world and the effects of our policies, to do the same for their progeny. This approach is consistent with Randall Baker's observation in Chapter 5 that environmental concerns are probably a better current emphasis than fossil energy exhaustion. It is also consistent with a gradual implementation of energy-environmental taxes, with their general but persistent impact. This is essentially a "rolling plan" approach in which each generation accomplishes what is politically feasible for the benefit of those whose welfare is of most concern to them.[12] The process itself will raise consciousness, which inevitably will (1) foster the development of a broader understanding of the consumption process as an important ingredient in policy formation (see Chapter 4) and (2) raise the question of our motivation for protecting future people as addressed by Norman Care in Chapter 7. Indeed, these issues are already topics of fundamental concern across the spectrum of those who are focused on conservation questions.

Full social cost pricing, as nearly as we are able to discern it and embody it in policy, may not be sufficient as an approach to a sustainable energy future. This is true even if there is strong resonance with other forces of social and cultural change. The position I have taken is that, for a variety of political, social, and economic reasons, it is a very promising place to begin.

Key Ideas in Chapter 6

- Providing for a sustainable energy and material future is inherently a process of continuous substitution and adaptation.
- There is no guarantee of, or clear decisive path to, the energy transformations required for a sustainable future.
- Reconciling economic growth and sustainable energy transformations requires a clear understanding that there is no one-to-one relationship between increases in income and wealth and the throughput of energy and material resources.
- The goal of policy is to "internalize" the full social costs of energy, material, and environmental resources in a manner that incorporates the needs of future generations.
- Under most circumstances, the gradual implementation of energy-environmental supplementary charges would be more effective and efficient in reducing energy consumption and environmental degradation than would particularistic forms of government regulation.

Notes

1. Donella Meadows et al., *The Limits to Growth* (New York: Universe, 1972).
2. Nicholas Georgescu-Roegen was a pioneer in the effort to bring the laws of thermodynamics to a position of central importance in economics, especially the economics of the environment and long-run growth and development. *Entropy Law and the Economic Process* (Cambridge: Harvard University Press, 1971).
3. This example and much of the flavor of the first half of this chapter are taken from a textbook that emphasizes a fundamental understanding, rather than the mechanics, of economic analysis. Paul Heyne, *The Economic Way of Thinking* (Upper Saddle River, N.J.: Prentice-Hall, 1997).

4. For an early and ultimately prophetic analysis of the problems and failures of centrally planned economies, see Frederick Von Hayek, *The Road to Serfdom* (Chicago: University of Chicago Press, 1944).
5. M. Bruce Johnson, "The Environmental Costs of Bureaucratic Governance: Theory and Cases," in John Baden and Richard Stroup, eds., *Bureaucracy vs. Environment* (Ann Arbor: University of Michigan Press, 1981), 218–19.
6. Heyne, *The Economic Way of Thinking,* 135–36.
7. Robert M. Solow, a Nobel laureate for his work in the theory of economic growth, has been an insightful leader in the application of economic principles to the sustainable growth problem. "The Economics of Resources or the Resources of Economics," *American Economic Review* (Proc.) 64 (May 1974): 1–14.
8. Harold Hotelling, "The Economics of Exhaustible Resources," *Journal of Political Economy* 39:2 (April 1931): 131–75.
9. Nathan Rosenberg, "Innovative Responses to Materials Shortages," *American Economic Review* (Proc.) 63 (May 1973): 111–18.
10. Tom Tietenberg, *Environmental and Natural Resource Economics* (New York: Harper Collins, 1996), chs. 14–19, esp. pp. 363 and 441.
11. William Stanley Jevons, *The Coal Question: An Inquiry Concerning the Progress of the Nation and the Probable Exhaustion of Our Coal Mines,* 3rd ed. (London: Macmillan, 1906; 1st ed., 1865).
12. Michael Kaganovich, "Rolling Planning: Optimality and Decentralization," *Journal of Economic Behavior and Organization* 29 (1996): 173–85.

Chapter 7

Protecting Future People: The Motivation Problem

NORMAN S. CARE

History is replete with tales of heroic sacrifice—for example, the yielding of a place in a lifeboat—and the news regularly reports rescue efforts made so that others may live or be free. But these acts usually have been in the face of a very immediate and recognizable evil, calamity, or threat. What if the threat is remote, uncertain, or not well understood?

Reducing energy consumption—and by extension, seeking more sustainable lifestyles—is fundamentally a moral issue. It makes sense only in an ethical framework. If one doesn't care about those who will inhabit the earth many generations in the future, why worry about sustainability? In this chapter, Norman Care moves from the cultural, political, and economic determinants of human behavior to the philosophical question of how current people can be motivated to follow the policies that morality requires—insofar as this can be identified—for the welfare of future people. Are we as individuals and members of a social order capable of developing an ethic that adequately protects people who are remote from us in both time and place? Our moral obligation to protect future people is assumed to be widely accepted. Even the differences between the "optimists" and the "pessimists" on the sustainability issue are not usually with regard to our moral obligation to the future. Rather, the disagreements are over what is feasible, desirable, and necessary to fulfill this obligation. Rejecting the views of the more extreme optimists—that existing social and economic structures are adequate to deal with our obligation—Care addresses the problem of motivating ordinary citizens to place burdens on themselves in order to enhance the welfare of others who are very remote from them in both time and place. This motivation is seen as a necessary precondition for an effective energy policy—especially in a democratic society.

—Editors' note

The Motivation Problem

Suppose we are independent policy makers for a free society. Our job is to formulate policies and put them before the larger community for approval. While we as individuals are members of the society we make policy for—we live in it and know it well—our appointments as independent policy makers give us enough security that we may be disinterested and impartial in our work. We are free to consider what morality requires in the matters raised by the various policy questions we take up, and we in fact attempt to devise satisfactory policy solutions for our society's problems that are in line with what morality requires—so far, of course, as we can make out what morality requires in particular cases. But, given the context in which we work, we are also mindful of the fact that ours is a free society and that in such a society it is best for the implementation of public policy to flow from general *support* for it on the part of the people, not just from, for example, legal coercion. Questions about the *availability* and *reliability* of such support recur in our work on policy questions. Finally, our recommendations to the larger community are respected: they will be considered seriously. In the past, our recommendations have often been accepted.

Our task at this point is to prepare morally principled policies concerning our society's legacy to "the world of the future."[1] These policies deal with many things but (for convenience here) principally with the distribution and conservation of natural resources and the control and impact of our physical plant on the environment. Suppose it becomes clear that the policies we find emerging from our work—the policies we develop under the guidance of our conception of what morality requires of us in this matter of our legacy to the world of the future—will not be easy to implement. We see that they call for serious sacrifice on the part of ordinary people.

Now, since we are policy makers for a free society and thus seek to recommend policies whose implementation can command general support, the fact that the morally principled policies we have developed make heavy demands for sacrifice by ordinary people is something we cannot pass over lightly. We recognize that the people of our society—our fellow citizens—have deep interests in the ways of life they labor (hard and honestly, in most cases) to build and maintain. These ways of life reflect our fellow citizens' different interpretations of the value of self-realization, and in some cases their ways of life reflect aspirations they have for those close to them, such as their children or even their friends and associates. Such interests, values, and aspirations seem legitimate to us (though we may quarrel with their interpretations or contents in some cases). Indeed, as members of society, we too have such interests, values, and aspirations. We may pause over the fact that the unbridled pursuit of self-realization can become corrupted into self-absorption. But that is an extreme, and the possibility of its occurrence does not impugn our thought that there are things that individual members of a free society owe themselves. People who take seriously their personal development and the well-being of those close to them are not thereby morally unserious. We realize that a sacrifice can be perceived as being too great relative to one's legitimate interests in one's own realization, one's aspirations for others, and the ways of life related to these interests and aspiration. We recognize that, to many of our fellow citizens, these considerations may seem to be *defenses* against the demands for sacrifice involved in the policies we are entertaining.

Nevertheless, after further work, let us suppose that we see that the policies that demand this sacrifice are indeed *required* by morality and that these familiar defenses we have considered are overridden. In that case, we (independent policy makers for a free society) face this question: what (apart from legal or other forms of coercion) is available and reliable in the way of support for the morally principled policies we now see we must recommend to the larger community? Alternatively: what in people can we draw upon to *motivate* them to follow the policies that morality requires regarding our legacy for the world of the future?

I will refer to this as the motivation problem for public policy regarding the future or, for short, as the future-generations motivation problem.

Understanding the Motivation Problem

How is this motivation problem to be understood? First, the problem at issue is not that of justifying morality to a thoroughgoing egoist, or what Hume called a "sensible knave."[2] The problem concerns the motivation of ordinary people, and they are not, in general (so I shall assume), egoists or knaves. Some may be, but not all or most.

Second, while it may be reasonably easy to generalize negatively that ordinary people are neither egoists nor knaves, it may be much more difficult to generalize positively about the motivation of ordinary people. What in fact moves people becomes at some point an empirical question, and this already suggests that there are limits to what philosophical work on this subject can achieve. A philosophical discussion of motivation available to support public policies guided by what morality requires for the world of the future will not, for example, yield statistical generalizations about the likelihood that the population of this or that nation-state can be moved by a given motivational factor, such as a sense of tradition or love of mankind. Nevertheless, there are certain aspects of this future-generations motivation problem that seem open to philosophical treatment. In what follows, I restrict my attention to some of these.

Third, I assume that the availability and reliability of relevant motivation is a condition of the acceptability of proposals for social policy in a free society. Supporting motivational factors are, of course, only some among the many types of considerations bearing upon the acceptability of policies. But they may be of special importance in the context of the development of policies concerning our legacy for the future. I assume that policies are unacceptable when it is known that rational persons of integrity cannot follow them or can follow them only with extraordinary difficulty such that they make demands nearly "exceeding the capacity of human nature."[3]

Besides, as a purely practical matter, suppose we understand that our ways of life—our individual activities and collective projects, together with our means of conducting and implementing them—jeopardize or seriously risk damaging the interests and prospects of future people. Suppose we understand further that, in these circumstances, morality requires substantial—even radical—changes in our ways of life. Then, given our understanding of these issues, it would surely *help* in the choice of social policies that are developed with the requirements of morality in mind to determine what motivation in support of such policies is available and reliable among the people expected to follow them.

Finally, let me mention that the motivation problem I wish to explore is not unique to the policy context in which we deal with what morality requires for the world of the future. It may be that the problem typically arises when public policy calls for sacrifice and the sacrifice is thought of as for the sake of people who are distant enough from us to be faceless and impersonal. Of course, the condition and plight (for example, the destitution) of *current* people who are distant from us can often be revealed to us through the gathering of particular facts. Individual current people, at any rate, can (in principle) become known to one another. But the same opportunity to know future people in similar detail is not available to us. This affects our motivation to do (if not our understanding of) what morality requires for the world of the future. But it does not restrict the motivation problem to the context in which we are concerned with acceptable policy for the future.

Concern for Generations to Come

Some theorists claim that people as we know them are not—and even cannot be expected to be—motivated by "a concern for generations to come" to act upon what morality requires for the world of the future. Robert L. Heilbroner, for example, offers the following overview of our situation:

> A crucial problem for the world of the future will be a concern for generations to come. . . . Humanity may react to the

> approach of environmental danger by indulging in a vast fling while it is still possible. . . . On what private, "rational" consideration, after all, should we make sacrifices now to ease the lot of generations whom we will never live to see?
>
> There is only one possible answer to this question. It lies in our capacity to form a collective bond of identity with those future generations. . . . *Indeed, it is the absence of just such a bond with the future that casts doubt on the ability of nation-states or socioeconomic orders to take now the measures needed to mitigate the problems of the future.*[4]

Heilbroner here raises the motivation problem I have in mind. It may be useful to be aware of his own estimate of its seriousness:

> There seems no hope for rapid changes in the human character traits that would have to be modified to bring about a peaceful, organized reorientation of life styles. . . . Therefore the outlook is for what we may call "convulsive change"—change forced upon us by external events rather than by conscious choice, by catastrophe rather than by calculation. . . . Nature will provide the checks, if foresight and "morality" do not. Thus in all likelihood we must brace ourselves for the consequences of which we have spoken—the risk of "wars of redistribution" or of "preemptive seizure," the rise of social tensions in the industrialized nations over the division of an ever more slow-growing or even diminishing product, and the prospect of a far more coercive exercise of national power as the means by which we will attempt to bring these disruptive processes under control. From that period of harsh adjustment, I can see no realistic escape. Rationalize as we will, stretch the figures as favorably as honesty will permit, we cannot reconcile the requirements for a lengthy continuation of the present rate of industrialization of the globe with the capacity of existing resources or the fragile biosphere to permit or to tolerate the effects of that industrialization. Nor is it easy to foresee a willing acquiescence of humankind, individually or through its existing social organizations, in the alterations of lifeways that foresight would dictate. If then, by the question "Is there hope for man?" we ask whether it is possible to meet the challenges of the future without the payment of a fearful price, the answer must be: No, there is no such hope.[5]

This is (to put it mildly) a troubling view. It rests directly upon a pessimistic claim about the motivation of ordinary people—that they do not, and perhaps (given their social conditioning) cannot, feel responsible for future people deeply enough to move them to "acquiesce" in policies demanding substantial "alterations" in their "lifeways."[6] If my assumption is fair—that the availability and reliability of relevant motivation is a condition of the acceptability of public policies in a free society—then this pessimistic motivational claim presents a matter of great importance to the design and adequacy of any policies for natural resources and our physical plant that we propose seriously for adoption in a heavily industrialized "affluent" society such as our own.

Let me ask what this concern for generations to come—which is said to be absent or weak—consists in and amounts to. In what follows, I explore several ways in which such a motivational factor may be interpreted.

Stipulations

Let us call people who live fifty generations after us "future people." Let us call ourselves "current people." In order to begin my exploration of the motivation problem, I must make three initial points. They are stipulations, and I will not attempt to argue for them.

The first point is that morality requires something of us regarding our legacy to the world of the future. Morality, as it were, speaks on this subject. What morality requires of us regarding future people is a full philosophical problem in its own right, and I do not take up that problem here. I only suggest some maxims that might stand as the "requirements of morality" regarding the world of the future.

We may begin by agreeing with Joel Feinberg's remark that "surely we owe it to future generations to pass on a world that is not a used up garbage heap."[7] Beyond this, let us suppose first that morality requires (among other things) an equal-opportunity maxim, according to which what we owe to future people is a world at least no worse off than the one we received. As Brian Barry puts it, "The overall range of opportunities open to successor generations should not be narrowed."[8] Second, it may require (among other things) what we might call a *custody maxim,* according to which current people are to regard themselves as "custodians rather than owners of the planet, and ought to pass it on in at least no worse shape than they found it in." This is to say, I take it, that morality may require a change in our attitudes (and relevant practices) toward the things of the earth. We may have to give up the view of the world as a cluster of resources that we can come to possess or own and, instead, adopt a view of the world as a cluster of resources of which we are temporary custodians or stewards.

I will not attempt to argue either for these maxims or against other candidates that may come to mind. Let us simply regard the maxims I have mentioned as plausible candidates to figure in an account of what morality requires of us in behalf of the world of the future.

The second point is that what morality requires of us is indeed what we (current people) would count as *sacrifice.* What exactly the sacrifice consists of I must leave indeterminate. Experts appear to make different estimates of what our natural resource situation is and of what shape our physical plant is in. The fact that current humankind is organized into nation-states is not helpful in this matter of information and estimates. Our approaches to the facts get cluttered, so to speak, with politics.

But, however far away we are from accurate information about our physical plant and resources, I share with many people a growing uneasiness about our situation. As laypersons relative to the technical matters sometimes involved, we have the impression that our resources are drying up or being squandered and that our physical plant is in disrepair or of the wrong kind to serve human needs in growing populations. All of this portends sacrifice. The future has the bleak look of a burden about it. The vocabulary of the policies we are asked to consider is that of "pulling back," "seeking alternatives," "using less," "curbing," "lowering," and "restricting." Those who see something objectionable in ways of life supported by affluence and considerable energy consumption may welcome the prospect. But many of us do not.

The third point is a distinction that allows the motivation problem to arise: it is one thing to understand what morality requires; it is another thing to be moved to do what morality requires. If understanding what morality requires operated in us in such a way that, in Kant's words, the actions that we recognize to be "objectively necessary" were also "subjectively necessary," then the motivation problem I have in mind would not arise. But—so I assume here, following Kant—this is not so.[9] There is no *tight* connection between understanding what morality requires and being moved to do what morality requires. It does not follow, of course, that understanding what morality requires has no motivating power. But it does follow from this distinction that an account which shows what morality requires in some category of cases does not *thereby* make clear (at least not directly) the motivation of people to do or act upon what morality requires in that category of cases.

Notes about Future People

Future people (those who live many generations after us) are faceless and impersonal to current people. The details that make people at least interesting to each other personally are missing. We do not know what their lifestyles are, what they stand for, whether they think much of us, or whether they are concerned about people who live many generations after them. Face-to-face encounters are out of the question.

We are not totally ignorant of what future people are like, of course. If (ex hypothesi) our descendants are still *people,* then we have now a body of salient facts about them. They will almost certainly have wants and needs and hopes and fears; they will almost certainly love and hate and perhaps even experience resentment and feel guilt. We may say, in overview fashion, that future people will have interests, and current people can do things now, so it seems, that affect those interests. We can do things now that hurt future people. We can damage the environment, for example, or use up finite resources without researching and providing for alternatives, thereby affecting how many future people there are, or make the circumstances in which future people live very difficult to bear.

These facts—that future people will be people and will have *interests* that current people can affect—may make it intelligible to say that people of both sorts (future and current) have rights and, in an abstract sense, belong to one moral community. If this can be said, perhaps it in turn helps show that indeed morality requires something of current people relative to future people. But, of course, even if such claims are intelligible, it does not follow that we should adopt them or base social policies upon them. Indeed, the whole idea that we and future people are members of a common moral community may seem strange. There is no mutual cooperation in such a community, and there are no exchanges "in kind"; as a consequence, this community is not characterized by mutual benefit or by joint participation in common activities in *any* familiar ways. The one "exchange" I can think of is that for the sacrifices we make in order to deliver an inhabitable globe to future people, they are grateful. They at least do not think little of us. But this "exchange" is hardly "in kind" and, in fact, it never reaches us.

In what follows I explore what these few notes about future people suggest concerning the motivation of current people to make the sacrifices required by morality in behalf of future people.

Motivation and Identity

It would be unrealistic for policy makers to rely on love or concern to motivate current people to do what morality requires for the world of the future. This is suggested by the faceless and impersonal character of future people. Future people cannot arouse love or concern in current people—at least not in the way other (typically nearby) current people can. Whatever may be the full explanation of this fact, it involves the lack of identity—the facelessness and impersonality—of future people. As I remarked above, the details that make people at least interesting to one another are missing in the case of future people, and the *capacity to interest* is a precondition of—or perhaps a constituent part of—the capacity to arouse in us such "motivational factors" as love or concern. Somehow being moved by love or concern for a person or a group of people involves having more of that person or those people before one than exists in the case of future people.

But perhaps this can be answered. We noted earlier that future people will have interests that we can know about now and affect now. Isn't that identity enough to make it possible that love, concern, or other strong feeling might be aroused? Perhaps we can love or have concern for future people insofar as they may be construed as interests we can affect now.

But I do not think this response is helpful. I do not think one can love or have concern for interests per se, and so it does not help make motivation in the form of love or concern flourish or even exist to quasi-reduce future people to their interests. What is at stake here *is* motivation—in particular, that motivation to do what morality requires that meets the terms of my earlier general characterization. The fact that future people have interests that we can affect now may indeed be relevant to the philosophical problem of determining whether or not we have responsibilities to future people—and, beyond that, to the philosophical problem of determining what, morally, is adequate provision for future people. But, again, what thus helps us to see or understand what morality requires is not thereby what motivates us to do what morality requires. And in any case, the problem of lack of identity remains. The interests of future people, construed as derived from their wants, needs, hopes, and fears, are radically indeterminate. Insofar as this is so (and risking the air of paradox), the interests of future people do not have the capacity to interest us. Therefore, so I suggest, a condition of their having the capacity to arouse love or concern or other strong feelings in us is not satisfied.

I said above that it would be unrealistic for policy makers to rely on love or concern to motivate current people to do what morality requires for the world of the future. Perhaps the term *unrealistic* is after all too weak. The discussion I have offered suggests that policy makers cannot rely on motivation in the form of love or concern, because motivation in this form is not available in the context of policy for the future. More exactly, what is not available is love or concern that has as a condition of its possibility (1) the capacity of its object to interest us and (2) which capacity involves the details of human identity. If a person were to profess or claim to be moved by love or concern for future people, or even for future people qua interests, I would have to say that something other than the love or concern I have indicated is being referred to.

In general, insofar as motivation to do what morality requires is tied to awareness of details in human identity, such motivation is not available to support widespread acquiescence in public policies meant to implement what morality requires for the world of the future.

Community Bonding and Reciprocation

A second form of motivation that is sometimes available to prompt us to do what morality requires may be labeled (for purposes of this discussion) *community bonding.* Even if this term is not ordinary, what I have in mind is familiar enough. We are commonly and often moved to act in ways that may be in line with principles (and hence, on occasion, with moral principles) by a regard for other persons that derives from their membership with us in a common community, association, enterprise, or project of some sort. At a minimum, motivation of this sort involves a *sense of belonging to some joint enterprise with others.* The motivational factor constituted by this sense of belonging may carry with it feeling-tones of solidarity, comradeship, loyalty, mutual confidence, and trust, or at least a sense of being on the same side. And the sense of belonging with others to a joint enterprise has a certain directionality to it. It is not indiscriminate in its objects. It selects those "others" who belong with oneself to that common enterprise. It picks out others who stand with one as comembers of some identifiable body, association, or community. In the event that one fails the association or violates the community or lets down the side, one may incur feelings of guilt or be dismayed or experience regret.

Just as there is a condition of the occurrence of motivation in the form of love or concern—by way of awareness of details in human identity—so there is a condition (probably many conditions) of the occurrence of motivation in the form of community bonding. I will call this condition *reciprocation,* by which I have in mind the exchange of ideas and conceptions of purposes that must be available to persons before they can be considered to stand as joint participants in a common project. We should distinguish this condition of reciprocation from that of prospective mutual benefit. Joint participation in common enterprises often has mutual benefit as its aim. But reciprocation is at a deeper level than the cooperation appropriate to mutual-benefit associations. There are communities, associations, enterprises, and projects that we participate in or even endure but that we do not view as directed toward mutual benefit. Participation in or endurance of these different associations and enterprises may arouse motivation in the form of community bonding, yet these various modes of joint participation cannot occur in the absence of the reciprocation I have in mind. People may find themselves in routines and regimens in which their behavior is controlled for them (for example, as unwitting victims of medical experimentation), and these routines and regimens may occur in the absence of reciprocation (and give rise to still different kinds of "motivational factors," such as resentment). But such modes of control or manipulation of behavior are not communities, enterprises, associations, or projects. The latter rest on a ground of exchange of ideas and interpretations of ends among their participants —an exchange that is sometimes one sided, no doubt, but nevertheless (to whatever minimal degree) present.

If we ask at this point whether motivation in the form of community bonding is available and reliable to support policies designed to implement what morality requires of us in behalf of the world of the future, I think the answer must be no. We (current people) and they (future people) are not positioned in such a way as to be able to reciprocate with each other concerning the constituent ideas and controlling aims of any associations or enterprises that we jointly participate in or endure. In the absence of the possibility of such reciprocation, I do not see how motivation in the form of community bonding can arise. The feeling-tones of solidarity or loyalty, or even the sense of being on the same side, are foreign to our relationship (whatever it is) to future people. Accordingly, the experiences of guilt, dismay, or regret—as they are known to us from our acquaintance with what it is to damage or destroy the associations we have with persons with whom we exchange ideas and interpretations of ends—cannot arise from our relationship to future people. If one should claim to feel guilt as a result of faults in one's conduct toward future people, I would have to say that the experience of guilt referred to is of a different kind from that which may be explained by reference to motivation in the form of community bonding.

In general, insofar as motivation to do what morality requires is tied to reciprocation with other persons concerning the ideas and aims of shared enterprises—as is so in the case of community bonding—then it is not available to support policies implementing what morality requires for the world of the future.

Extended Shared-Fate Motivation

Finally, let me consider a form of motivation that seems very different from love or concern and community bonding. In those cases, the motivation was construed as grounded either in the identity of other persons or in reciprocation with them concerning the ideas and aims of joint enterprises. The motivation I wish to touch on now is more abstract in character. I will call it *extended* or *unbounded shared-fate motivation.* I have in mind our sense of common humanity (if we have it and to whatever extent we have it), which involves at some level the notion that in a very general way human beings as such "belong together" or are "in life together" *irrespective* of differences in time and location.

Let me offer a series of thoughts, in no special order, about motivation of this kind. First, we need some account or picture of what extended shared-fate motivation consists of or at least what it is like. The following passage from Rawls's *A Theory of Justice* is suggestive, I think, of the possibility of the form of motivation I have in mind:

> Individuals in their role as citizens with a full understanding of the content of the principles of justice may be moved to act upon them largely because of their bonds to particular persons and an attachment to their own society. Once a morality of principles is accepted, however, moral attitudes are no longer connected solely with the well-being and approval of particular individuals and groups, but are shaped by a conception of right chosen irrespective of these contingencies. Our moral sentiments display an independence from the accidental circumstances of the world. [10]

What attracts my attention in this passage is its suggestion that motivation may be grounded in an idea independently of the details of human identity or what I have called reciprocation between persons. Given my general approach in this chapter, let us ask what such an idea would have to be *of* in order to serve as a ground for motivation-supporting public policies operating in behalf of the world of the future.

So far as I can see, the most straightforward candidate for an idea that could serve this function—analogously to the role of detailed identity in the case of love or concern and to the role of reciprocation in the case of community bonding—is the idea of a community of persons who may be at any temporal or social location and who nevertheless construe themselves as "being in life together" or "sharing fate" according to the contents of an appropriate conception of what morality requires of the members of such a community. We must note at once that the idea of community thus brought before the mind is unusual in important ways:

- It is an idea of a community in which not all or even very many of the members can reciprocate with one another—that is, engage in an exchange with one another over the ideas and purposes of their community. No real *joint* decision making, no matter to what extent representation is employed, can take place in it.
- It is an idea of a community in which not all or even very many of the members have—in principle—particularity for one another. Most of its members are, and must remain, faceless and impersonal to one another.

- It is an idea of a community whose membership is *unbounded* in all the ordinary ways. Thus, it has no national or even geographical limits, and its membership extends into the future indefinitely, if not infinitely.

Once the idea that might ground extended shared-fate motivation is sketched in this way, we may ask: can an idea of this sort in fact be motivating among persons as we know them? That is, can we not only conceive of a community of this sort, and of ourselves as members of it, but also imagine ourselves developing a sense of belonging to it such that that *sense* might be available and reliable in support of policies meant to implement what morality requires for the world of the future? It is one thing, we may suppose, for us to be able to form the conception of humankind as an unbounded (yet) shared-fate community; it is another thing for such a conception to arouse in us, for example, *affection for* the community.

Allow me to comment on the prospect of an unbounded community of mostly nonreciprocating persons who are lacking in identity for one another being motivating among people as we know them.

First, it is worth mentioning that reference to this idea is an occasional part of ordinary moral discourse—although this occasional use of the idea may not indicate serious appeal to the content I have suggested the idea might have. I have in mind, in the context of discussions of policy for the future, appeals to common humanity in the form of appeals to the fact that *persons,* after all, may be hurt by what we do now.

Second, we may think of the development of certain forms of motivation as describable by very general psychological laws linking their emergence to institutional structures and practices.[11] Thus, given a society or social setting of a certain sort, together with persons' natural sentiments, we may expect certain forms of motivation to be aroused in the human beings who live in that society or social setting. To think in this way about the development of motivation leads us in the present case to ask whether our society is such that we may reasonably expect the people who have their lives in it to develop extended shared-fate motivation.

When we raise this latter question, my own estimate is that our society is not one that lends itself to the cultivation of extended shared-fate motivation. To show why this is so would be a large project, and I will not attempt it here. It would first involve assessment of the structures, institutions, practices, and moral ideology characteristic of our society. Next, it would require treatment of the further questions of what sorts of social structures would lend themselves to the development in persons of a sense of belonging to an unbounded community of human beings that could be strong enough to provide reliable support for public policies implementing what morality requires for the world of the future. Without meaning to beg important questions, I suspect that this large project would find that certain elements in the makeup of our society—for example, the lack of Good Samaritan rules in our legal system, the self-interested psychology of competitive appropriation fostered by our economic life, the emphasis on self-realization in our educational system, and the prizing of immediate sensibility so powerfully supported by everyday commerce and culture—operate *against* the development in us of extended shared-fate motivation.

In general, I think extended shared-fate motivation is intelligible—we can imagine its presence in persons—and I think it "fits" our thoughts concerning what motivation would have to be like to be serviceable in the context of policies implementing what morality requires for the world of the future. But there is nevertheless serious empirical doubt, I think, about its availability and reliability as support for public policies in a society such as ours. Given the influence on us of certain dominating institutional elements in the makeup of society, plus our current received moral ideology, it may be that motivation of the unbounded shared-fate sort is somehow beyond us—or too difficult for very many of us to develop—at this time. The earlier discussion of love or concern and of community bonding brings to our attention familiar features of our nature, but extended shared-fate motivation is, while not unknown to us, nevertheless not similarly familiar.

On Novel Ideas

Reflection on all of this leads to pessimism in one's estimates of our chances of offering a decent legacy to future generations. Are there other sorts of motivation we should explore?

In what follows, I discuss a form of motivation that is both familiar and different in kind from certain of the factors mentioned above. It is different from what "love or concern" and "community bonding" seem to involve, although it is suggested by the structural feature of extended shared-fate motivation whereby that form of motivation rested ultimately on a very general *conception* of community. I have in mind what has traditionally been called the *power of ideas,* and what I will here call the *motivational power of novel ideas.* By "novel ideas," I mean broad conceptualizations of the world we experience that are—relative to our current conceptualizations—unusual, intellectually odd or challenging, and sometimes different in what they suggest for attitudes and action. By the "motivational power" of these ideas, I refer to the power these new ideas can develop to guide our thoughts and thus our actions. It is the power these ideas have when they "take" in us—that is, become "operative" in our ways of thinking about the world—and, in effect, constitute a part of "common knowledge" for us.

It seems to me uncontroversial that novel ideas can come to have motivational power. Here are two examples of what I mean.

First is an egalitarian moral idea—the idea that all persons have certain rights (for example, "natural rights" or "basic rights") and are thus morally on a par with one another. This idea is both widespread and familiar; it expresses an ideal that is profound for many people; and it moves some people to conduct their lives in certain ways—in some cases, in ways involving protest and radical political action. At one time, this idea was novel, but over time it caught on and became motivating. Now, it is to many of us not novel at all; we are used to it, and it is so entrenched in us that its way of structuring our thoughts about how persons are to be understood seems to us a matter of common knowledge. It is even operative among people whose governments do not take it seriously or even acknowledge it, and it survives even when there is controversy over what the basic rights are.

The second example is an interpretive idea about the conditions of life for many people: the idea that poverty is not necessarily a natural state that some members of society must be in but is instead a function of conventions, such as economic systems influencing political and legal systems. This idea of poverty as a conventional rather than natural state was novel at one time, but over time it "took"—that is, it caught on and became motivating—and now it is to many of us not novel at all, even though we differ in our views of how we should respond to the poverty around us. If the sky is cloudy today, that is (in most cases) a natural fact, and while one may not like the fact, one is not morally offended by it. But if one becomes aware of extensive poverty in one's local or national community, or even in the world community, one's reactions are different: if one attends to the situation and learns about its details, one very likely will be offended by it. There is something wrong with extensive, deep, life-stifling poverty. It ought not to exist.

Of course, even if poverty ought not to exist and one believes that this is so in a way that involves the idea that poverty is a function of conventions rather than natural, it does not follow that one will *do* anything about the poverty that exists—for example, work to ameliorate or eradicate it. But the problem I have in mind is not the one prompted by the fact that people sometimes do not do what they believe they ought to do. (Sometimes people have reasons they find compelling for not doing what they believe they ought to do.) The motivation problem I have in mind rests at an earlier point. It does not have to do with moving from understanding the right thing to doing the right thing. Rather, it involves how candidate novel ideas come to have standing in one's understanding of the right *so that* the familiar problem of choosing whether or not to do what one recognizes to be right can then arise in a way that involves reference to those ideas.

How is it then that a new idea might not only intrigue us but also come to *count* with us such that we think differently about certain elements in our experience and perhaps, as a result, actually alter our conduct or ways of life respecting those elements? If certain ideas (for example, that all persons have basic rights or that poverty is a function of conventions) tend to regulate our thoughts—and thus operate as standing presumptions for our conduct—so that our *not* acting in line with their normative content requires an explanation, how do candidate novel ideas come to be regulative for us? I should say that my interest in this question (and this discussion) is not in the analysis or assessment of ideas—that is, with the "contents" of a conception or with the arguments, evidence, or justifications available to it. It is no part of my view that the questions of analysis and assessment are unimportant or unanswerable. But I suggest that the "motivation problem" I have in mind remains, in particular cases, even when the analytic and assessment questions are satisfactorily responded to. Even when a broad conceptualization is clear and the arguments for it are strong, it doesn't *follow* that it will "take." Also—perhaps unfortunately—when a conceptualization is not clear or when the arguments for it are not strong, it sometimes "takes" in a person nevertheless. My attention here goes not to what makes a novel idea a good or a bad idea; it goes instead to what is involved in an idea's becoming a functioning part of one's practical outlook in some domain of experience.

Of course the ideas I mentioned—about persons and poverty—are hardly novel for most of us. Let me next explore the future-generations motivation problem by discussing a less-familiar idea, one important in environmental ethics. While this idea may be familiar to theorists in environmental ethics, my impression is that it is not (yet) an idea that is operative among people generally.

An Environmentalist Idea

The idea in question belongs to the branch of environmentalism that goes beyond appeals to prudence in our use of the environment and its resources for human purposes. This branch of environmentalism holds that the environmental crisis we face in fact calls into question our sense of our own standing in the larger whole of nature itself. On this view, while we may indeed develop and pursue public and personal strategies for cutting back our consumerist ways of life as well as increase our efforts in technological research and the development of alternative sources of energy, we must first rethink our place among the natural creatures, objects, and systems that make up our environment. Indeed, we must move away from our received anthropocentric sense of ourselves as the sole moral agents in the natural world to the cultivation of an "ecological conscience" that treats human beings as members *with* other members (that is, other natural creatures, objects, and systems) of a larger whole: the biotic community itself. A general prudence regarding the environment may permit us to continue to think of ourselves in the familiar terms of the anthropocentric ethical traditions we have inherited. But that response, so it is argued by the environmentalism I have in mind, is shallow and is not morally adequate to the serious crisis we face. The development of a genuinely "ecological conscience" is, according to this environmentalism, the required next step in the ethical evolution of humankind.

This notion of ecological conscience involves relatively novel ideas, I think, but it is worth noting that ecological conscience itself is not totally unfamiliar to us. (It is not found only in the moral and cultural sensibility of "native peoples.") Thoreau's *Walden* is a classic of the Western literature we think of as American, and Thoreau's combination book and journal elucidates the notion of ecological conscience in moving detail.[12] Thoreau's love for Walden is not just "appreciation of nature" in the way in which from time to time most of us are moved by natural phenomena. This love for Walden has wrapped up in it Thoreau's responses to certain philosophical problems about how one is to live and what kind of person one is to be. To some extent, Thoreau's responses are negative: they push us away from concern for the externals that tend to overwhelm our lives, such as property, material goods, power, and reputation. On the positive side, Thoreau drives us in on ourselves. The "individuality" that Thoreau is famous for encouraging turns out not to be a matter of the idiosyncrasies or features of temperament and possessions by which one differentiates oneself from others. It is, rather, the essential core of "significant and vital experiences" we all share.[13] Thoreau's "experiment in simple living" is "to live deliberately . . . to drive life into a corner, and reduce it to its lowest terms."[14] Among these significant and vital experiences, for Thoreau, is a sense of membership in the natural world involving a recognition of values in nature that is, values *in* (not "projected into") the various items and systems that make up the natural order. Thoreau's critique of our life "in society" is that we have raised ourselves above nature, subordinated it to our purposes, and made it instrumental to our material interests. Thoreau objects to the reduction of nature to "our property," and he admonishes us "to enjoy the land, but own it not."[15] His thought is that, in the larger view, we belong *with* the other creatures and elements he finds in his life at Walden under "the greater Benefactor and intelligence."[16] Our smaller view, whereby we give ourselves special standing in the order of things, he regards as the corruption of vanity.[17]

This idea of a form of membership in nature that involves the recognition of value *in* the environment and the items and systems that make it up is, to the environmentalism I am sketching, of central importance to the achievement of depth in our thinking about environmental crisis. Already in the foreword to *A Sand County Almanac,* Aldo Leopold writes: "We abuse land because we regard it as a commodity belonging to us."[18] When, instead, we regard ourselves as parts of the natural order, other parts of which also have value, then "we see land as a community to which we belong" and "we may begin to use it with love and respect."[19] In *Philosophy Gone Wild* and *Environmental Ethics* and *Conserving Natural Value* by Holmes Rolston III, the idea of value *in* nature is again a main object of attention, though now the philosophically problematic aspects of the idea are faced directly.[20] Rolston indicates that to attempt to develop this idea is to "swim against the stream of a long-standing paradigm that conceives of value as [merely] a product of human interest satisfaction."[21] It is nevertheless his view (echoing Thoreau and Leopold) that nature is not to be reduced to "instrumental value" but is, instead, to be considered a reservoir of "intrinsic value" and that recognition of this fact is central to an adequate response to the environmental crisis.

Notes on the Power of Novel Ideas

What can we glean from this small discussion of novel ideas—two examples of ideas once novel but no longer so; one example of an idea now (relatively) novel but, as it were, a serious contender for a place in thought and action? What is it for novel ideas to "take" in us in such a way that thought and action are influenced? Could the idea of "ecological conscience"—an idea that radically reconceives our place in nature—come to have motivational power in us?

I offer some notes that respond to these questions, though my notes do not reach to the techniques and strategies involved, for example, in politics or in coercive manipulations of persons, no matter how noble or benevolent their ends. Rather, I address what it is for novel ideas to "take" at (I think) a more basic level. In any case, I would regret discovering that a positive result for future generations rests upon only manipulation or external coercion of persons.

Notice first that for novel ideas to be motivating they apparently must be interesting—intellectually interesting—in some way or another. They must, so to speak, catch the attention of the mind. Consider when classical theorists claim that persons have certain basic rights (whether government agrees or not), or the social thinker urges us to see that poverty is not natural but is instead a

Figure 7.1. The power of novel ideas.

function of conventions, or the environmentalist finds intrinsic value in the natural order—in all of these cases, there is some conceptual maneuvering occurring. Reconceptualizations of familiar items (persons, poverty, the environment) are proposed, and those of us concerned to understand the world and not just walk around in it become intellectually fascinated.

But that, of course, is not enough for "motivational power." An idea can be interesting in its manner of contrasting with current or received ideas but still fail to be motivational. What else is involved?

I notice further, in the examples I have worked with, that in each case, beyond the reconceptualizing of familiar items, there is a *normative proposal*—that is, a proposal bearing on our attitudes and, through them, on our actions. This seems to me germane to the issue of how an idea can have motivational power, for the point of a normative proposal is to urge us to adopt a new way of *valuing* what the novel idea entails. Typically, the reconceptualization carries with it implications concerning the importance of its subject and how it is to be treated. Thus, the egalitarian basic rights theorist is saying that persons—all of them—are worthy of respect, whether they are tall or short, smart or dumb, religious or not, and that they ought to be treated accordingly. The social thinker is saying that poverty is objectionable as well as able to be eliminated and that we ought to think about and act toward it accordingly. Certainly, the environmentalists are saying that the natural world and its constituent parts have intrinsic value, and thus worth independent of human-projected value, and that we ought to think about and act toward them accordingly.

I think the normative element in the novel idea is important: the candidate novel idea offers not only a different reading of some item in or part of experience but also one that is charged with the energy of recommendation or demand or objection. This aspect of the situation raises interesting and difficult questions about states of mind in both the "source" and the "recipient" of the candidate idea insofar as it is possible for one to receive a novel idea *as* recommendation or demand (or threat or opportunity) when no such thing is intended. But I will not pursue these complexities here. The more intriguing point for this discussion is that even when a novel idea is intellectually interesting and contains a normative proposal (correctly received, say), this may not be sufficient for the idea to "take" in one. Notoriously, normative proposals, even in the form of justified ethical demands (for example, principled calls for respect for items of this or that kind), generally do not necessarily move us into thought and action in line with their contents. I ask again: what else is involved when a novel idea comes to have power in us? In what follows, I offer some speculative notes regarding this question. I do not have a full answer to it, though I believe that the question itself has a place in the development of a comprehensive understanding of how ideas relate to action.

The first note is reminiscent of a theme of Rousseau's in *A Discourse on the Origin of Inequality* (1755).[22] Over and beyond the intellectual-interest elements and the normative element in a novel idea, it *helps* that novel idea be or become motivating if the idea "strikes a primitive chord" in us. There are, of course, different ways of striking primitive chords. In one case, we may imagine that the idea penetrates the conceptual layers we have taken on through education and under the pressures of various social and cultural forces and that it reaches to some original experience we have had. Perhaps most of us have had our share of experiences in which we have "felt" ourselves to be part of nature. The environmentalist novel idea may function, in part, to recall one's experiences of this kind (if any), and insofar as the idea functions in this way, it might somehow gain what I have been calling motivational power. It might draw such power from the power of the original experience in one—supposing, of course, that the original experience itself had salience in a relevant positive or negative fashion.

In another sort of case, the primitive-chord striking may be rather different.

Here, we imagine the situation in which one's life is filled with failure and frustration and one's experience is confusing and filled with moral-emotional dissonance. It seems in such a case that, within one's life, routine ways of thinking are ineffective and that one's agency is diminished. In these circumstances, novel ideas can sometimes provide a combination of illumination and guidance that enables the recovery of agency; when this is so, it helps the novel ideas "take" in one, even though the "taking" is not guaranteed. I think that, in fact, such recovery linked to novel ideas occurs, although it remains mysterious to me how the "gap" between the availability of the novel idea and its "taking" so that recovery is facilitated is to be described. Internally, the novel idea seems to sort experiences for one in helpful ways—its "taking" might be reflected in one's sense of finally having "insights" into one's experiences—and the recovery of agency comes to seem possible to one. One is able to "move on" in one's life.

There may be an "externalist" version of this sort of case, too. It *helps* a novel idea become motivating if it speaks to or addresses (through its normative element) external facts or situations that are both troubling (for example, they offend moral sensibility) and also objects of repeated attention. For example, if it is a novel idea that our responsibility to aid others in distress extends far beyond those close to us (such as family members or friends) to suffering members of humankind in general, then it helps this novel idea gain in motivational power when one gives repeated attention to the troubling facts of destitution and inequality in levels of life of members of humankind in today's flawed world community. It appears that as we become more intimate with certain troubling facts, a novel idea that addresses them, even if initially strange, comes to seem serious, familiar, relevant, and finally (sometimes) even right. Again, in circumstances of these kinds, a novel idea may sort the troubling experiences and in doing so afford guidance for action.

Second, suppose a given novel idea urges one to respect items in a range different from the range to whose items one is used to giving respect. Then one is more nearly ready to take this proposal seriously if one has already learned to give respect to certain other items that are also outside the range of those items to which one is used to according respect. For example, I suspect it becomes easier to be moved to respect items in the environment, and ecosystems of various dimensions, if one has already begun to offer respect to, say, animals as well as human beings. Perhaps, indeed, if one is used to offering respect to persons, and then the relevant idea "takes" such that one also offers respect to animals, one might then find motivating the idea that still other items in the environment (for example, wilderness tracts or, more generally, ecosystems of various dimensions) are worthy of respect.

The third note looks at the other end of the motivational process. I think that if a novel idea in fact comes to have power for one, then one cannot, as it were, "go back." If the idea that poverty is a function of conventions "takes" in me, then the earlier notion—that poverty is a natural state for some people—is lost to me. Similarly, if the environmentalist's idea of ecological conscience "takes" in me, I think the older idea—the natural order and its items are merely resources or instruments for me—is lost to me. Of course, one might intellectually study or review older ideas. But, motivationally, a novel idea that "takes" seems to cancel the *power* of the older idea. Later, one might move to another novel idea in some domain of experience, but in general, when a novel idea becomes motivating, it diminishes and perhaps terminates the power of the idea it replaces or supplants.

I should mention that I do not hold that novel ideas always motivate (when they do) because they make one "feel good" or they otherwise "serve one's interest." I am reminded here of Aldo Leopold's sad remark: "One of the penalties of an ecological education is that one lives alone in a world of wounds."[23] Indeed, to take seriously the environmentalist's idea of ecological conscience in today's world would be to bring into view many "wounds" to our planet, but this (ordinarily) would not lead to one's "feeling good," and whether or not it served one's interests would depend on the circumstances and the nature of those interests. When a novel idea "takes," this is not always due to the standard motivational factors recognized by consumerist ideologies and interests-oriented politics.

The fourth note makes apparent that my observations here are indeed speculative. Suppose that the environmentalist's idea that nature has intrinsic value has power for one. Perhaps this is so because it combines intellectual interest and a normative proposal and somehow reaches into one, through the conceptual layering, to make contact with original experiences one has had. Suppose further that the appeal of this idea, when first encountered, was immediate and easy for one. (Perhaps it was put forward by very able teachers.)

In contrast, in other cases this general subject brings to mind, an idea's coming to have motivational power is much more labor intensive and struggle filled. (The "taking" comes harder, so to speak.) Consider, for example, the case of the novel ideas offered to the (alleged) hopeless drunk "bottoming out" after, say, a thirty-year drinking history. There are cases in which, through the good work of Alcoholics Anonymous (AA), self-destructing human beings have been coaxed back to sobriety and then have successfully moved on to vigorous lives involving families and careers. Now, "ideas" are not the only things involved in the typical AA recovery story, of course. The recovering alcoholic in AA is surrounded and supported by people and is invited to move slowly and patiently through daily study and discussion of AA's twelve-step recovery program (the "novel idea," for the purposes of this note). This is not the place to explain this program or discuss this fascinating sort of case in detail.[24] What is important for this part of my discussion is that the ideas that AA offers its new members are often not motivating *at first;* they *become* so as time passes and as the new member, typically suffering the residual effects of his or her past and greatly fearful of (and at once attracted by) the prospect of returning to drink, takes slow steps among the members and with the AA program. In some cases, over time, the novel ideas, which were perhaps even hated in the beginning, somehow come to have power. The dynamics of this process needs to be better understood. It involves pain and suffering in a way that is not so in the imagined case in which the environmentalist's idea "takes." I cannot explore the process here, however; I only acknowledge its existence and in that way record it among these notes.

Conclusion

I end with some perhaps obvious caveats. First is a recognition of the difficulty of theorizing about what I have described as an idea's coming to have "motivational power." It seems to be a fact—one taken into account in the previous discussion—that an idea that "takes" in one person may not "take" in another, despite, for example, the achievement of a similar level in the idea's clarity and justification for both persons. One might worry, then, that the "taking" of an idea in a person is "subjective" in a sense that prevents its being an object of a theory or philosophical account; perhaps there is no list of "conditions" that a theory or account might provide such that when those conditions are satisfied, the candidate idea "takes." In this view, the ideas that are operational in us do not come to be so (merely) by meeting conditions of clarity or justification or by satisfying other rational or cognitive tests, and it is inferred that this "coming to be operational" is outside the ken of theory.

My thought here is that it does not follow from this worry about subjectivity that nothing illuminating can be said about what it is for an idea to "take" when it does. Many different things undoubtedly are involved in an idea's "taking" or "not taking" in a person—things as disparate as social pressures, misunderstandings of facts, vulnerability to self-deception or gullibility, need for stability, self-confidence, even chemical imbalances in the brain. Still, the facts that there is such complexity and that an idea that "takes" in one person may not "take" in another do not themselves render an idea's "taking" opaque to theory or philosophical description. Besides, apart from how far a theory might be helpful in understanding these facts (and illuminating the complexity), the initial point that the "taking" of ideas is logically beyond satisfaction of conditions of clarity and justification itself has a place in a philosophical account of how ideas relate to thought and action. There is no reason yet to infer that this "taking" cannot be an object of theory.

The second caveat is that I hold no special brief for "novel ideas" in general. I recognize that not all novel ideas are morally admirable, even as I also recognize (especially in retrospect) that some novel ideas are meritorious and should come to have motivational power. Clearly, the sheer fact that an idea involves a reconceptualization of experience plus a normative proposal does not mean that that idea *should* "take." We must be on guard with novel ideas, as with ideas generally. Still, apart from how we attempt to clarify and thus understand the ideas we find interesting and then assess them in terms of cogency, internal consistency, evidence, and justification, there is, I think, a question about what it is for them to come to have motivational power that deserves attention. Nothing I have said suggests that a novel idea's coming to be motivating is either easy or simple; nothing I have said means to minimize uncritically the power (or merit) of older ideas we possess and live by; nothing I have said limits what may count as "motivational factors" affecting thought and action. In some cases, it is a good thing for a novel idea to come to have motivational power; in other cases, it is not a good thing. In cases of either kind, what it is for a novel idea to come to have motivational power is a philosophical puzzle.

Key Ideas in Chapter 7

- Sustainability is fundamentally a moral issue.
- The central question addressed here concerns how people can be motivated to do what morality requires regarding the rights of future generations.
- The usual forms of motivation are seen as inadequate for the protection of people who are remote in place and time.
- Development of a widely accepted "environmental ethic" or "ecological conscience" is suggested as a required next step in the social evolution of humankind.

Notes

This chapter is a condensed and edited version of Chapter 6 of *Decent People,* by Norman S. Care (Rowman & Littlefield, 2000).

1. The phrase is Robert L. Heilbroner's in *An Inquiry into the Human Prospect* (New York: Norton, 1975), 114.

2. See David Hume, *An Enquiry Concerning the Principles of Morals,* edited by Tom L. Beauchamp (Oxford; New York: Oxford University Press, 1998).
3. For discussion of "the strains of commitment," see John Rawls, *A Theory of Justice* (Cambridge: Belknap Press of Harvard University Press, 1971), 145, 176–77, 423.
4. Heilbroner, *An Inquiry into the Human Prospect,* 114–15 (Heilbroner's italics).
5. Heilbroner, *An Inquiry into the Human Prospect,* 131, 132, 135–36.
6. Heilbroner's pessimistic motivational claim may be controversial, of course. See Joel Feinberg's remarks: "I shall assume . . . that it is psychologically possible for us to care about our remote descendants, that many of us do in fact care, and indeed that we ought to care." Feinberg, "The Rights of Animals and Unborn Generations," in Richard A. Wasserstrom, ed., *Today's Moral Problems,* 2nd ed. (New York: Macmillan, 1979), 581.
7. Feinberg, "The Rights of Animals," 598.
8. The equal-opportunity maxim is discussed by Brian Barry in "Circumstances of Justice and Future Generations," in R. I. Sikora and Brian Barry, eds., *Obligations to Future Generations* (Philadelphia: Temple University Press, 1978), 242–44. The custody maxim is endorsed though not discussed by Barry in "Justice between Generations," in PP. M. S. Hacker and Joseph Raz, eds., *Law, Morality and Society* (Oxford: Clarendon, 1977), 284. Barry remarks at the end of the latter essay that these maxims might form the minimum content of the "new ethics" that some theorists call for regarding our obligations to future people.
9. Immanuel Kant, *Foundations of the Metaphysics of Morals* (1785), trans. and intro. Lewis White Beck (Indianapolis: Bobbs-Merrill, 1959), sec. 2, p. 29.
10. Rawls, *A Theory of Justice,* 475.
11. Rawls, *A Theory of Justice,* 491.
12. Henry David Thoreau, *Walden* (1854), intro. Norman Holmes Pearson (New York: Rinehart, 1948).
13. Thoreau, *Walden,* 274.
14. Thoreau, *Walden,* 74.
15. Thoreau, *Walden,* 173.
16. Thoreau, *Walden,* 277.
17. Thoreau, *Walden,* 275.
18. Aldo Leopold, *A Sand County Almanac* (1949) (New York: Ballantine, 1970), xviii.
19. Leopold, *A Sand County Almanac,* xviii–xix.
20. Holmes Rolston III, *Philosophy Gone Wild* (Buffalo, N.Y.: Prometheus, 1986); *Environmental Ethics* (Philadelphia: Temple University Press, 1988); *Conserving Natural Values* (New York: Columbia University Press, 1994).
21. Rolston, *Philosophy Gone Wild,* 73.
22. In *The Social Contract and Discourses,* trans. and intro. G. D. H. Cole (New York: Everyman, 1950).
23. Leopold, *A Sand County Almanac,* 197.
24. I offer explanation and discussion in my *Living with One's Past: Personal Fates and Moral Pain* (Lanham, Md.: Rowman & Littlefield, 1996).

Conclusion

Few problems facing humankind today suggest more strongly the need for an interdisciplinary perspective than the interrelated problems of energy, the environment, and sustainability. We have approached these problems from a variety of disciplines and made a concerted effort to integrate them into a coherent whole.

We have had two major concerns. First: what determines future energy availability, its projected use, and the nature and consequences of the energy conversion required to meet basic and acquired human needs? Second: what are the complex problems of our moral, social, political, and economic responses to the challenges posed by energy sustainability? We have stressed both the inexorable nature of the depletion of specific physical resources and the uncertainty about the consequences of their exhaustion for human well-being.

Chapters 1 to 3 addressed the fundamental laws of nature and the physical and environmental limitations on energy use. Chapters 4 to 7 focused on the human and societal implications, perceptions, and responses. The fundamental questions concern, first, the sources of energy for a world in which consumption is rapidly growing, and second, the effects on the environment of growing energy use. Can we develop concepts of "quality of life" and "consumption" that incorporate environmental quality and the continuous availability of the energy required for their support? Our quest for answers raises the basic scientific, cultural, political, economic, and philosophical questions that are addressed in this work.

Energy that sustains all life and life-related activities on earth is finite in supply, and human uses of energy tend to convert useful forms to useless ones. These energy flows are subject to laws of nature that cannot be altered by human intervention. Moreover, they limit our ability to provide adequate energy to meet human needs and aspirations for a satisfactory worldwide standard of living that can be sustained into the foreseeable future and beyond. Meeting the world's quality-of-life goals for all citizens during the twenty-first century will require population stabilization, the development of new energy technologies, substantial improvements in the efficiency of energy use, and significant changes in lifestyles. A magic bullet in the form of a new energy source unknown to science today is not likely.

Meeting the challenge of providing an adequate energy base is only the front half of the energy problem. The other half is how to achieve this without doing intolerable and irreparable damage to the environment. There is no such thing as a *clean* energy source—all energy transformations have environmental consequences. This is an inescapable implication of the laws of thermodynamics. Consequently, environmental impacts rather than energy shortages may be the most important limiting factors on energy use in the twenty-first century.

The impact of human beings on the environment is due mainly to the enormous amount of energy we use compared to other species. Therefore, a better understanding of the cultural origins and patterns of energy consumption is crucial to achieving energy and environmental sustainability. Such an understanding could provide policy makers with new tools for linking broad social issues to policies aimed at conserving energy and the environment. Of particular importance are the cultural factors that change human wants into needs. Whether the psychological, social, and cultural changes needed for sustainability can occur in time to save the planet from unacceptable and irreversible environmental degradation resulting from overpopulation and overconsumption is an open question.

In the past, it was easy to point the finger at such "polluters" as steel mills, auto plants, and municipal sewage systems, but now it is clear that the ultimate polluters are *us* in our cars, *us* in our homes, and *us* (whether rich or poor) in our general consumption patterns. None of us like to point the finger at ourselves as the beneficiary of pollution resulting from underpriced goods and services and as the potential cause of deprivation for future generations.

The excuse that something is "not economical" is not citing some immutable law of nature. Rather, it is, importantly, an observation that reflects the current state of our values as well as some obvious market and policy imperfections. These imperfections reflect the fact that energy and other resources are currently being delivered to the consumer at a price that fails to reflect the full social and economic costs of their use. The external costs that are not considered in market pricing are not infinite but neither are they zero. So, if alternative energy sources cannot become competitive, then it is not necessarily because they are "not economical" but because we, as a society, are wedded to a short-term perspective and to ranking current consumption over prudent, or even self-interested, regard for present and future consequences.

Policy in a democratic society reflects the preferences and values of that soci-

Figure 8.1. Ignored alarms.

ety, and there needs to be widespread public understanding and concern before anything can be done. An energy policy, or any policy for that matter, is something that results from a groundswell of public concern that arises from reality or perceptions of reality (the nuclear "risk," for instance). Thus, the role of the individual is important, even though it may take a specific event to focus public concern. Most people derive their information from knowledge and "signals," particularly economic market signals. At present, there are too few of these that suggest a crisis—current or in the making—with regard to either the exhaustion of fossil fuels or the existence of environmental and ecological threats. For this reason, it is important to develop a broader and longer-term public understanding of the nature of consumption and the consequences of all forms of the energy conversion required to sustain it.

One of our goals in writing this book was to reconcile the scientific and economic world-views—to explain how growth toward an acceptable global level of consumption can be obtained and sustained continuously in a finite world. There are two aspects to this reconciliation. One is the principle that economic growth consists not in increasing the production of *things* but in the production of *wealth,* where wealth is whatever people value. Whereas material things contribute to wealth, there is no fixed relationship between growth in wealth and the use of material and energy. This is particularly relevant for the growth of affluent economies, and it will become more applicable to developing economies as their per capita incomes rise above subsistence levels. Sustainable economic growth requires continuous scientific discovery and technological development to make possible the substitution of wealth in the form of reproducible capital for wealth in the form of exhaustible resources. How effectively this process can continue is the sustainability question.

The other aspect of the reconciliation concerns the impact of economic development on ecological systems ("the environment," broadly defined). For the highly developed economies, our ancestors arguably bequeathed to us in the form of skills and technology embodied in reproducible capital more than they took from us in the form of depletable and nonrenewable natural resources. This substitution obviously has its limits, however, especially when it involves the depletion of our ecological capital. Although most people in developed countries live substantially better lives than their ancestors did in terms of the comforts and health benefits of modernity, it is unlikely that the energy-intensive lifestyles in developed countries could be ecologically sustainable on a global scale.

We have focused on the important questions but are not pretending to know all the answers—no one does. Our primary goal has been to change perceptions and attitudes, not to give specific policy recommendations. Nevertheless, some general goals suggested by this work for the world in the twenty-first century are clear. They include the following:

- development of new technologies for cleaner and more efficient use of fossil fuels to meet increasing world energy demand
- accelerated development of alternative energy resources to complement and ultimately replace fossil fuels—including the technology that enables them to meet acceptable standards of environmental impact and safety
- reduction in per capita energy use in the developed world
- increased standards of living in the developing world with less than proportionate increases in the energy required for their sustenance
- stabilization of world population with a smaller per capita energy use than has been required historically in developed countries.

Our main message is that achieving these goals is a complex, interdisciplinary problem that cannot be solved without a basic understanding of both the laws of nature that constrain our options and the fundamental moral, cultural, economic, and political principles that determine how humans behave.

Natural science keeps us in touch with the laws of nature (the first and second laws of thermodynamics) and gives us our basic understanding of what realistic options we have for meeting energy demand in the twenty-first century in environmentally acceptable ways. The key insight of science is that long-term energy sustainability will require development of new technologies for utilizing solar energy in all its forms—direct solar radiation as well as wind and other forms of renewable energy that are driven by the sun. Science also tells us that solar and wind energy (the two most promising renewables), while widespread and long lasting, are of very low concentration compared to fossil fuel energy and, therefore, have limits of their own because of the large collection areas required. Finally, science tells us that a "magic bullet"—a new energy source unknown to science today that would permanently solve our energy problems—is not likely. There is no purely technological fix; science and technology will not be able to keep up with world energy demand in the twenty-first century and beyond at acceptable environmental costs if growth in energy demand continues at the present rate. The energy resources simply are not there, and the environmental impacts would likely be unacceptable if they were. Thus, meeting long-term energy demand in the twenty-first century will require worldwide stabilization of population and energy consumption per capita.

The social sciences and humanities give us insights into how this stabilization might come about. Difficult choices and tradeoffs will have to be made, and they will be culturally dependent. Because the competition for energy resources and the environmental impacts of energy use are global, understanding the cultural factors that affect energy use is crucial to finding workable solutions to the energy problems in different parts of the world. Economic and political realities will also affect the feasibility of different approaches to energy sustainability. We can make things worse by not understanding the complex interactions of culture, economics, and policy. There will be many different solutions to energy problems, each within a different historical, economic, political, and cultural context.

At the core of all sustainability issues are fundamental moral questions concerning our obligations to future generations. While some sense of moral responsibility for future generations is fairly universal, it is not obvious, given uncertainties, missing signals, and short-run orientation, how people can be motivated to do what morality requires regarding the rights of future generations. The usual forms of motivation are weakened and likely to fail with regard to people who are remote in place and time. A new and widely accepted "environmental ethic" or "ecological conscience" may be the next step in the social evolution of humankind required for long-term energy and environmental sustainability.

Truly *long-term* sustainability will require fundamental changes over time in social and cultural institutions and practices. Love and concern for our children and grandchildren can provide strong *short-term* motivation for us to look ahead two to three generations and move toward a longer term conservation ethic. It is up to each generation, therefore, to be well informed, to head in the right direction, to learn from experience, to take advantage of opportunities, and then to *pass the baton* to the next generation. Working toward a sustainable future is best viewed as a continuous process of adaptation to changing circumstances. The slower the required rate of change, the easier the process of adaptation will be for future generations. This is a compelling argument for stabilizing world population and, sooner rather than later, reducing the inefficient and wasteful uses of natural resources.

Animal species seek to dominate their habitat. Ecological balance is a balance of aggressive reproduction. When they are too successful, species can destroy their habitat—and, thus, themselves. Only humans, in the interest of nurturing their own species, have the capacity, the institutions, and the insights to change the *form* of their dominance to one based on ecological conscience, extended shared fate, enlightened self-interest, and other guiding principles. For implementation, these principles require a broad public understanding of what is at stake, recognition of both our limits and our opportunities, and the development of political will.

Appendix 1

Possible Energy Sources on Earth

GLENN YOUNG AND JOHN SHEFFIELD

With regard to potential energy sources for the world, one might ask the following questions:

- What sets the limits on energy availability?
- Are there energy sources we are not currently using?

In considering these questions, it should be remembered that an energy *source* delivers energy. Electricity, hydrogen, batteries, and fuel cells are examples of energy *carriers*—some other form of energy gave them their energy. For example, hydrogen may be formed by the electrolysis of water using energy from an electrical power plant. In turn, the power plant used energy from coal, oil, natural gas, or uranium, or from solar energy evaporating water from the sea and depositing it as rain in the hills, where its potential energy can be released as it returns to the sea.

Simply put, the answers are as follows:

- Energy is limited primarily to sources of energy existing on the earth today and energy input from the sun.
- It is probable that the only source of energy remaining to be exploited is that from the fusion of light elements.

How do we know this? Energy derives from the actions of forces, and high-energy physicists tell us that there are three fundamental types: gravitational, electro-weak, and strong nuclear forces. Back in time, as the universe emerged from the Big Bang, these forces, in combination and succession, determined the form of our universe:

- Gravity was the dominant force up to about 10^{-32} second after the start when the temperature was about 10^{23} electron-volts (eV) (1 eV is equivalent to about 11,600°K). The highest energy cosmic rays are observed near such energies.

- Around 10^{-11} second after the Big Bang, when the temperature was about 10^{12} eV, the electro-weak force split into two parts—the weak interaction and the electromagnetic force. Effects at such energies can be probed with the most powerful accelerators. The weak force is involved in the fusion of hydrogen in the sun to make the heavier hydrogen isotope deuterium.
- Strong forces played a particularly important role at around 10^9 eV (1 billion eV) and much less than one second after the Big Bang, producing protons through the fusion of quarks. At around one second and a temperature of about 10^7 eV (10 million eV), these forces also produced the light elements deuterium, tritium, and helium by the fusion of protons.
- The production of the heavier elements, also by fusion, occurred and is still occurring within the dense material of stars at temperatures typically less than 10^6 eV (1 million eV). Fission energy is released when some of the heavier elements, such as uranium and plutonium, are split to make lighter elements. Standard solar burning leads to elements up to iron, while the really heavy elements (for example, lead and uranium) have to be made during a supernova explosion to have any significant quantity—not something easy to do technologically!
- Atomic forces involving the electromagnetic interactions of electrons and charged ions (which constitute atoms and molecules) become important at less than 1 eV (11,600°K). They are involved in ordinary chemical reactions.

Thus, the fundamental origin of all energy resources on earth is predominantly nuclear fusion in the core of the sun and other stars. This nuclear fusion produces:

- the elements, whose chemical rearrangement can release chemical energy—for example, carbon + oxygen —> carbon dioxide—in an exothermic process (one that gives off heat)
- the heavy elements, such as uranium, which can be fissioned in an exothermic process
- sunlight, which provides direct energy supporting plant growth and renewable biomass energy resources, provides heat and electricity directly, generates the wind and waves, and raises water against the force of gravity, which then falls as rain on higher ground and whose gravitational potential energy then can be used to do work or make electricity.

In addition, the heat produced by nuclear radioactive decay in the center of the earth provides geothermal energy, and the kinetic energy of the moon, as it orbits the earth, provides tidal energy through the gravitational force.

The one known form of energy that we do not yet use directly is nuclear energy from the fusion of the lighter elements. While there may be other sources of energy that are realizable, such additional exothermic sources have not been identified. If they exist, they would be related to those forces that were important closer to the Big Bang and at higher temperatures than 10^7 eV (80 billion°C). It seems improbable that they can be of practical use, since even investigating such a region cannot be done efficiently today. Therefore, it seems sensible to consider only those sources of energy that have been identified as economic or potentially economic.

Energy derived from sunlight is viewed as renewable because it is so long lasting—the sun is expected to continue shining for several billion years. To summarize, these renewable resources include:

- heat and electricity from direct sunlight
- biomass energy from agricultural crop and animal waste, forestry wastes, and energy crops
- wave power
- wind power.

Geothermal energy is limited by heat flow from the earth's core to regions of accessible useful temperature. Chemical energy from the existing fossil resources—oil, gas, coal, hydrates, shale oil, tar, peat—are nonrenewable.

Nuclear energy is available from fissionable heavy elements, such as uranium-235. In addition, the more abundant uranium-238 and thorium may be converted into the exothermically fissionable elements plutonium-239 and uranium-233, respectively. While there are large amounts of uranium available on the earth, it is a nonrenewable resource.

The light elements deuterium (D) and lithium are a potential source of fusion energy. Tritium (T) may be produced from lithium by neutron bombardment for use in the most readily achievable fusion scenario, D-T fusion, which occurs at a temperature of about 100 million°C. Deuterium (heavy hydrogen) at a fractional abundance of 1/6,500 of all hydrogen is, in effect, limitless in the oceans, but reasonably priced lithium is not a limitless resource. Deuterium-deuterium (D-D) fusion is possible in principle, but it requires a higher temperature (400 million°C). The fusion of deuterium produces helium-3 and tritium, which may be recycled to make the deuterium fusion more effective. A more speculative option is to mine helium-3 on the moon and use it in the D-He3 cycle. Some consideration has been given to more exotic fuel cycles, such as fusing protons and boron, but they are even harder to achieve.

Appendix 2

Energy Units

There is a great assortment of units for measuring familiar physical quantities, such as distance, time, mass, weight, and volume. The choice depends on country, convention, and the size of the quantity being measured. The British system (inch, foot, mile, second, ounce, pound, tonne, quart, gallon, and so forth) is used in Great Britain and the United States; the metric system (centimeter, meter, kilometer, second, gram, kilogram, cubic centimeter, cubic meter, liter, and so forth) is used in most other countries. Small units (inch, centimeter) are used to measure small quantities (for example, the size of a book) and large units (miles, kilometers) are used to measure large quantities (for example, the distance from New York to Chicago).

Similarly, many units exist to measure the quantity of energy, the choice depending on the form of energy (electric, gas, coal, food, and forth), convention, and convenience. This means that conversions often have to be made from one system to another, just as currency has to be exchanged when traveling from one country to another.

It is a convention when discussing world energy use to define all energy in terms of the "equivalent energy" in a metric ton (tonne) of oil, because oil is such a dominant part of the world's energy supply today. One tonne of oil (toe) contains about 42 billion joules (J) of energy, while a tonne of high-quality coal contains about 29 billion J. In electricity use, it is common practice to speak in terms of kilowatt-hours (kWh) of energy, in which 1 watt (W) of power is a joule per second, so 1 kWh equals 3.6 million J, and 1 toe is equivalent to about 11,670 kWh.

Given below are the units used in Chapter 2, some other commonly used energy units, factors for converting one unit to another, and a graphical comparison of various unit sizes.

Energy Units Used in Chapter 2

Unit	Equivalent
1 tonne of oil equivalent (toe)	42 GJ (gigajoules) = 42 x 10^9 joules
1 tonne of coal equivalent (tce)	29.3 GJ
1000 cubic meters of natural gas	36 GJ
1 tonne of natural gas liquids	46 GJ
1 tonne fuel wood	0.38 toe
1 tonne uranium (current reactor, open cycle)	10,000 toe
1 tonne uranium (breeder)	500,000 toe
1 barrel of oil (159 liters = 42 U.S. gallons)	0.136 tonnes
1 BTU	1055 J
1 Quad	10^{15} BTU
1 Quad	25.12 Mtoe
1 calorie	4.187 J
1 watt (W)	1 joule/second

Numerical Prefixes

Prefix	Equivalent
Exa	10^{18} (quintillion)
Peta	10^{15} (quadrillion)
Tera (T)	10^{12} (trillion)
Giga (G)	10^9 (billion)
Mega (M)	10^6 (million)
Kilo (k)	10^3 (thousand)

Abbreviations

Unit	Abbreviation
Hour	h
Year	a (annum)

Weight Equivalents

1 metric ton (tonne) = 2,205 lbs = 2.205 U.S. tons
1 U.S. ton = 2,000 lbs = 0.907 tonne

Energy Conversion Factors

	Quad	Mtoe	Mbbl	KW-hr	kcal	Btu	cal	ft-lb	joule
1 Quadrillion Btu	1	25.13	162.31	2.93×10^{11}	2.52×10^{14}	1.00×10^{15}	2.52×10^{17}	7.78×10^{17}	1.06×10^{18}
1 Megatonne Oil Equivalent	0.0398	1	6.46	1.17×10^{10}	1.00×10^{13}	3.98×10^{13}	1.00×10^{16}	3.10×10^{16}	4.20×10^{16}
1 Million Barrels Oil Equivalent	6.16×10^{-3}	0.155	1	1.81×10^{9}	1.55×10^{12}	6.16×10^{12}	1.55×10^{15}	4.79×10^{15}	6.50×10^{15}
1 Kilowatt-Hour	3.41×10^{-12}	8.57×10^{-11}	5.54×10^{-10}	1	8.60×10^{2}	3.41×10^{3}	8.60×10^{6}	2.65×10^{6}	3.60×10^{6}
1 Kilocalorie (food calorie)	3.97×10^{-15}	9.96×10^{-14}	6.44×10^{-13}	1.16×10^{-3}	1	3.97	1,000	3.09×10^{3}	4.18×10^{3}
1 British Thermal Unit	1.00×10^{-15}	2.51×10^{-14}	1.62×10^{-13}	2.93×10^{-4}	0.252	1	252	7.78×10^{2}	1.06×10^{3}
1 Calorie	3.97×10^{-18}	9.96×10^{-17}	6.44×10^{-16}	1.16×10^{-6}	1.00×10^{-3}	3.97×10^{-3}	1	3.09	4.18
1 Foot-Pound	1.28×10^{-18}	3.23×10^{-17}	2.09×10^{-16}	3.77×10^{-7}	3.24×10^{-4}	1.28×10^{-3}	0.324	1	1.356
1 Joule	9.48×10^{-19}	2.38×10^{-17}	1.54×10^{-16}	2.78×10^{-7}	2.39×10^{-4}	9.48×10^{-4}	0.239	0.737	1

Comparison of Energy Unit Sizes

Unit	Joules	Equivalent
1 joule	1	0.737 fl•lb
1 ft-lb	1.36	Energy required to raise a 1-pound weight a vertical distance of 1 foot
1 cal	4.18	Energy required to raise the temperature of 1 gram of water 1 degree Centigrade
1 Btu	1.06E+03	Energy required to raise the temperature of 1 pound of water 1 degree Fahrenheit
1 kcal	4.18E+03	1,000 calories = 1 food calorie
1 kw-hr	3.60E+06	Energy required to run a 1,000-watt electric heater for one hour
1 Mbbl	6.50E+15	Energy required to heat 1 million Indiana houses in the winter for 1 week
1 Mtoe	4.20E+16	Equivalent to 6.46 million barrels of oil; one barrel (bbl) = 42 gallons
1 Quad	1.06E+18	The U.S. consumed 80 Quads of energy and the world 322 Quads in 1990

Large Units (Quad, Mtoe, Mbbl): national and world energy use.
Medium Units: (kw-hr, kcal, Btu): home heating and air conditioning; food.
Small Units: (cal, ft-lb, joule): moderate physical activity; small electrical appliances.

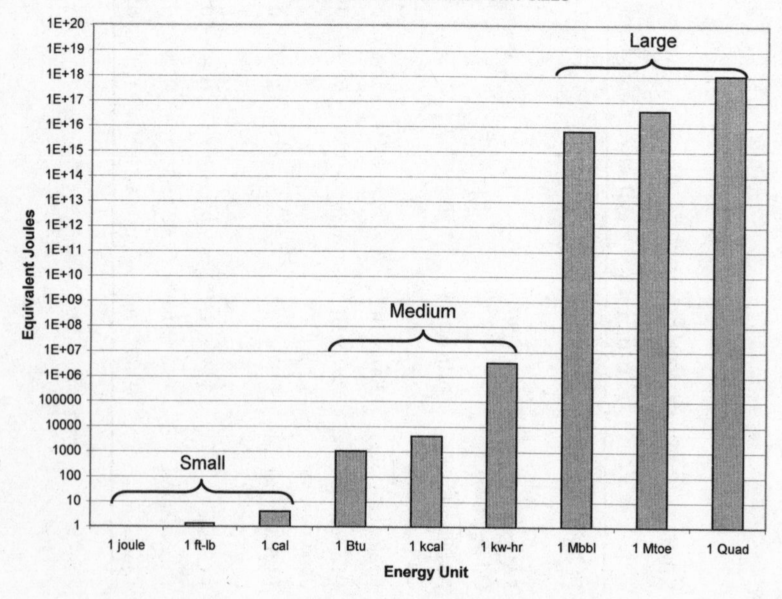

Appendix 3

A Formula to Relate Population Growth to Per Capita Energy Use

An empirical formula for the population growth rate [G(%)] that exhibits the key features of the dynamic plots in Figure 2.3 is described in Sheffield ("Population Growth"). Its variables may be interpreted as indicating the different effects of (1) a changing standard of living, interpreted as the useful annual energy use per capita (E in toe/cap•a) (that is, the part of the energy that is not wasted); and (2) cultural factors introduced through a dependence of the formula on the annual energy per capita for which the growth rate is zero (E_c).

$$G(\%) = (E_c - E)/(1.6 \times E^{0.38}), \text{ for } 0.05 \leq E \leq E_c. \quad (1)$$

The annual per capita energy value E_c (toe/cap•a) is the level at which the fertility rate equals the replacement rate and the population in an area stops growing. A value for E_c of 2 to 3 toe/cap•a, for energy use today, fits the trends in Figure 2.3 and is consistent with the fertility curve shown by Goldemberg and Johansson *(Energy as an Instrument for Socio-Economic Development).*

In terms of raising the standard of living, and thereby facilitating other changes, it is the energy used wisely that counts. Therefore, in relating energy use per capita to population growth rate, Sheffield considers the useful energy per capita per annum as the surrogate for standard-of-living influences on population growth rate, taking the year 2000 as the base. Thus, if the efficiency of energy use in 2020 improves by a factor of h from the level in 2000, the amount of energy needed to achieve a given goal will be reduced by $1/h$. In other words, if $h = 1.25$, 1 toe of energy in 2020 is as useful as 1.25 toe was in 2000. With the effective energy (E_e) normalized to the year 2000 as $E_e = h \times E$, the equation becomes:

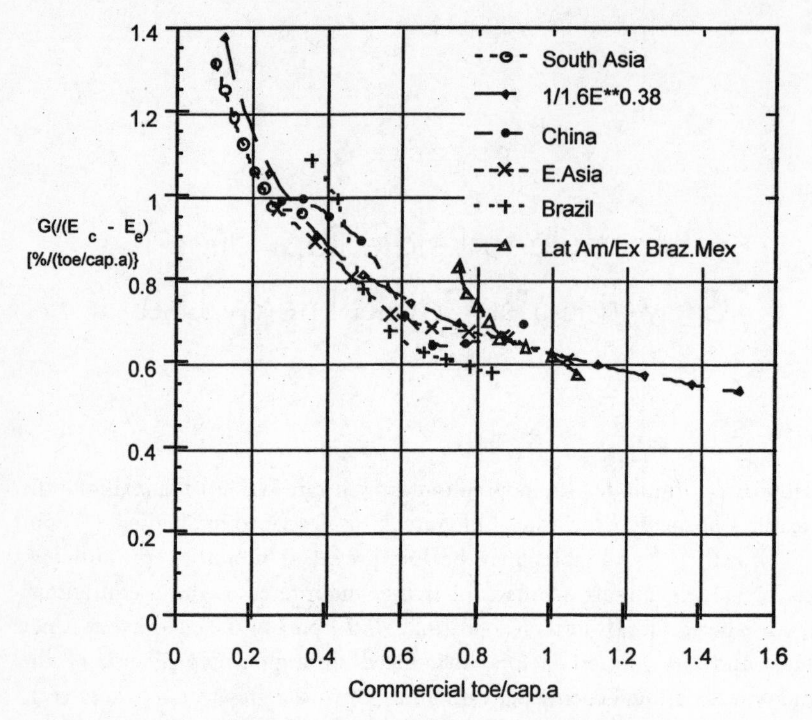

Figure A3.1. Plot of population growth rate, normalized to $E_c - E_e$, against annual commercial energy use per capita, 1965–1992.

Source: John Sheffield, not published, ORNL.

$$G(\%) = (E_c - E_e)/(1.6 \times E_e^{\ 0.38}), \text{ for } 0.05 \le E_e \le E_c \text{ toe/cap} \cdot \text{a}. \quad (2)$$

The good fit of this simple formula to the existing data is shown in Figure A3.1. The use of a different value for E_c (the energy value for which the growth rate is zero) for each area reflects the observed (cultural) differences between, say, China and countries of South Asia, which use relatively less annual energy per capita for a given population growth rate than Latin America.

The population growth rates projected by this approach for each part of the developing world are shown in Figure A3.2. For this hypothetical case, it was assumed that the overall efficiency of energy use improves steadily, starting in 2000, so that energy use is reduced by 37.5 percent in 2050. By 2100, the efficiency improves by 75 percent. Beyond 2100, there is a slower rise to a 91 percent improvement by 2150 and 100 percent by 2200 (that is, by 2200, twice as much useful energy is extracted per toe as in 2000). Potential efficiency improvements are discussed above.

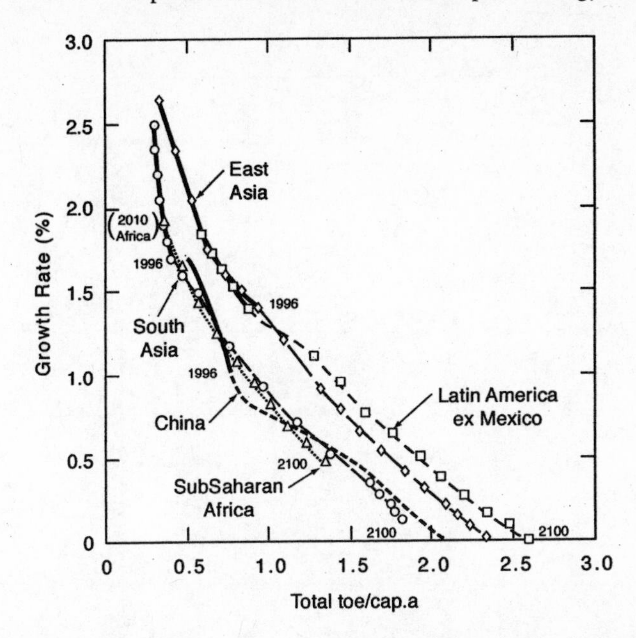

Figure A3.2. Hypothetical case: the population growth rate for developing regions of the world, historical data and projections up to 2100, as a function of total annual per capita energy use.

Source: John Sheffield, not published, ORNL.

For the developed world—North America, OECD Europe, FSU, CEE, and the Pacific OECD—this case assumes that energy efficiency gains will be used to raise standards of living so that energy per capita remains at the year 2000 level. The world's energy use rises to about 19,000 Mtoe/a in 2100. If efficiency gains were used to reduce energy use in the developed world, the world would use about 16,000 Mtoe in 2100.

For this projection, it was also assumed that cultural changes would steadily lower the population growth rate. This effect was implemented by lowering the "cultural" factor E_c:

For an effective energy use (E_e) of	E_c was lowered by
<1.0 toe/cap•a	1.0% per decade
1.0–1.5 toe/cap•a	1.8% per decade
>1.5 toe/cap•a	2.5% per decade

Appendix 4

List of Acronyms

CAC	command and control
CEE	Central and Eastern Europe
CH&P	combined heat and power
DOE–EERE	U.S. Department of Energy–Energy Efficiency and Renewable Energy
DOE–EIA	U.S. Department of Energy–Energy Information Agency
EPA	U.S. Environmental Protection Agency
FSU	Former Soviet Union
HDI	Human Development Index
IASA	International Institute for Applied Systems Analysis
IEA	International Energy Agency
IPCC	International Panel on Climate Change
ITER	International Thermonuclear Experimental Reactor
NIF	U.S. National Ignition Facility
OECD	Organization for Economic Cooperation and Development
OPEC	Organization of Petroleum Exporting Countries
PNGV	Partnership for a New Generation of Vehicles
SUV	sport utility vehicle
WEC	World Energy Council

About the Contributors

ANDREW BACHER is a professor of physics at Indiana University-Bloomington. His research interests include intermediate energy nuclear physics, nuclear astrophysics, and the applications of nuclear physics to biology and medicine. He has published more than ninety papers in professional journals, magazines, and book chapters, and is coauthor of several books. His interest in energy and environmental physics stems from his teaching of undergraduate courses on these topics.

RANDALL BAKER is director of international programs and professor in the School of Public and Environmental Affairs at Indiana University. His areas of specialization include international environmental policy, environmental management in the tropics, and the civil services in transitional economies. He is the author of *Environmental Management in the Tropics* (1992), *Summer in the Balkans* (1994), and *Comparative Public Management* (1995). He has recently completed a study entitled *Environmental Law and Policy in the United States and the European Union* and a work on the comparison on the transition to democracy in both right- and left-wing authoritarian countries (2000). Baker's principle professional interest is in establishing new schools and departments of public affairs overseas.

ROBERT BENT is professor emeritus of physics at Indiana University-Bloomington. A Fellow of the American Physical Society, he was a Guggenheim Fellow at Oxford University in 1962–63 and associate director of the Indiana University Cyclotron Facility in 1983–84. He has published over sixty research papers on experimental nuclear physics in professional journals and edited one book, *Pion Production and Absorption in Nuclei* (1982). Dr. Bent's interest in energy and environmental physics evolved in part from his undergraduate teaching in those areas.

MARY ELLEN BROWN is director of the Institute for Advanced Study and professor of folklore and ethnomusicology at Indiana University. A historical ethnographer, she works primarily with eighteenth and nineteenth century Scottish materials, with particular attention to cultural politics.

NORMAN S. CARE was a professor in the Department of Philosophy at Oberlin College for thirty-five years. His areas of interest in teaching and writing were moral theory, moral psychology, political philosophy, environmental ethics, medical ethics, and philosophy of art. He is the author of *On Sharing Fate* (1987) and coeditor of a number of collections, and has published essays and reviews in journals in philosophy, law, and education and in magazines of social comment. He is also the author of *Living with One's Past: Personal Fates and Moral Pain* (1996) and of *Decent People* (2000).

LEE H. HAMILTON is director of the Woodrow Wilson International Center for Scholars in Washington, D.C., and director of the Center on Congress at Indiana University. He served as a U.S. Representative from Indiana from 1965–99. During his years in Congress, he chaired numerous committees, including the Committee on Foreign Affairs (now the Committee on International Relations), the Permanent Select Committee on Intelligence, the Joint Economic Committee, and the Joint Economic Committee on the Organization of Congress. He serves today on numerous boards and committees.

RUSSELL LEE is distinguished research and development staff member for science and technology policy at Oak Ridge National Laboratory. He has over twenty-five years experience in the analysis of energy supply and demand, energy–economic modeling, and analysis of the environmental externalities associated with energy production and use. He has written over 150 published papers, books, book chapters, and technical reports, including the eight-volume series *Estimating Externalities of Fuel Cycles*, published by McGraw-Hill/UDI. Dr. Lee is winner of a Martin Marietta Awards Night Award for Technical Achievement in 1994, the Best Report in Geography Award from the Association of American Geographers in 1995, and a Lockheed Martin Awards Night Award in 1999. He led the teams that developed a number of models which the U.S. Department of Energy has used for the *Annual Energy Outlook* and for other publications.

LLOYD ORR is professor emeritus of economics and former chair of the economics department at Indiana University-Bloomington. In 1978–79 he was in residence as visiting professor of economics at the University of Kentucky and in 1983–84 as Culpeper Fellow at Oberlin College. His research work, presented in over fifty articles, research monographs, book chapters, and seminar papers, is concentrated on environmental economics and the economics of risk and safety regulation.

JOHN SHEFFIELD is director of the Joint Institute for Energy and Environment at the University of Tennessee and co-director of a joint institute of the Oak Ridge National Laboratory and the Kurchatov Institute in Moscow. He is involved in research in the energy and environmental areas. Dr. Sheffield's particular fields of study are future energy demand and alternative energy sources for the world, biomass and fusion energy, opportunities to improve the handling of animal manure, and air pollution from transportation. His Ph.D is from London University and he is also the author of the book *Plasma Scattering of Electromagnetic Radiation* (Academic Press [1975] and Atomizdat, Moscow [1978]).

WILLIAM Z. SHETTER is professor emeritus of Germanic Studies at Indiana University, with a specialization in the fields of linguistics and the language and culture of the Netherlands. Besides foreign languages, his other lifelong passion has been in graphic art, where he has had experience in various media. These two sides have been happily joined more than once in the chance to illustrate his own books. His concern for the matters raised in this book drew him into the discussions from the outset.

IAN THOMAS carried out research in superconducting magnetometry at the Open University, UK (where he obtained a Ph.D.) and at Vanderbilt University in Nashville, TN. He has taught math and physics at all levels from middle school to adult education, and is currently the physics instructor at Columbia Independent School in Columbia, MO. Thomas also publishes articles, instructional and documentary videos, and web sites for academic and generalist audiences on science and healthcare topics.

RICHARD R. WILK is chair of the department of anthropology at Indiana University, and president of the Society for Economic Anthropology. He has done ethnographic research on household decision making, energy consumption, and consumer culture in Belize, Ghana, and the United States. He has published extensively on the importance of consumer culture in understanding global environmental problems.

GLENN R. YOUNG is group leader for the Nuclear Physics Experimental Program in the physics division of Oak Ridge National Laboratory. He obtained his B.A. in physics and mathematics at the University of Tennessee (1973) and his Ph.D in nuclear physics from the Massachusetts Institute of Technology (1977). He spent one year at M.I.T. as a Chaim Weizmann Fellow and then accepted a Eugene P. Wigner Fellowship with Oak Ridge National Laboratory in 1978, where he has been since. His current research interests are in the area of relativistic collisions of heavy atomic nuclei to search for a novel state of matter with deconfined quarks and gluons.

Index

Island Press Board of Directors

A Garland Series

OUTSTANDING DISSERTATIONS IN THE

FINE ARTS